Operation, Analysis, and Design of Signalized Intersections

A Module for the Introductory Course in Transportation Engineering

First Edition

Michael Kyte
Maria Tribelhorn

Operation, Analysis, and Design of Signalized Intersections
A Module for the Introductory Course in Transportation Engineering

First Edition

By
Michael Kyte
Maria Tribelhorn

Cover Design by Beth Case

ISBN-13: 978-1500204365
ISBN-10: 1500204366

2014.07.03

PREFACE

In 2009, as a result of a conference on transportation education held in Portland, Oregon, a group of university transportation faculty initiated the National Transportation Curriculum Project (NTCP). The purpose of this project is to develop a new set of curricular materials for the introductory course in transportation engineering that are based on principles of active learning, focusing on a clear set of learning objectives, and with connections to how practicing professionals conduct their work. The project has spawned a number of technical papers, workshops, research projects, and new curricular materials in topic areas commonly taught in this course, including traffic operations, geometric design, and transportation planning, among others.

As part of the NTCP, in 2011, we began a study of what students learned (or didn't learn) about signalized intersections in the introductory transportation course. We analyzed examination results, homework assignments, and interviews with students to identify concepts that were particularly difficult for them. We reviewed existing textbooks commonly used in the introductory course to identify how concepts and topics related to signalized intersections were treated. We also reviewed standard guidebooks used by transportation professionals to identify concepts that would be appropriate to include in a new curriculum.

From these data, we developed a set of learning objectives that would govern the development of a new curriculum on signalized intersections that could be taught in the introductory course in transportation engineering. This module is the first product of that work. Additional materials such as critical thinking exercises, classroom activities, and design projects are being developed and will be distributed at a later date. Other related materials such as ranking tasks have been developed by other research teams.

The sequence in which concepts are presented in this module, and the way in which more complex ideas build on the more fundamental ones, was based on our study of student learning in the introductory course. The development of each concept leads to an element in the culminating activity: the design and evaluation of a signal timing plan in section 9. For example, to complete step 1 of the design process, the student must learn about the sequencing and control of movements, presented in section 3 of this module. But to determine split times, step 6 of the design process, four concepts must be learned including flow (section 2), sequencing and control of movements (section 3), sufficiency of capacity (section 6), and cycle length and splits (section 8). Depending on the pace desired by the instructor, this material can be covered in 9 to 12 class periods.

Before they begin their university studies, most students have experience with traffic signals, as drivers, pedestrians and bicycle riders. One of the tasks of the introductory course in transportation engineering is to portray the traffic signal control system in a way that connects with these experiences. The challenge is to reveal the system in a simple enough way to allow the student "in the door," but to include enough complexity so that this process of learning about signalized intersections is both challenging and rewarding.

We have approached the process of developing this module with the following guidelines:

- Focusing on the automobile user and pretimed signal operation allows the student to learn about fundamental principles of a signalized intersection, while laying the foundation for future courses that address other users (pedestrians, bicycle riders, public transit operators) and more advanced traffic control schemes such as actuated control, coordinated signal systems, and adaptive control.

- Queuing models are presented as a way of learning about the fundamentals of traffic flow at a signalized intersection. A graphical approach is taken so that students can see how flow profile diagrams, cumulative vehicle diagrams, and queue accumulation polygons are powerful representations of the operation and performance of a signalized intersection.

- Only those equations that students can apply with some degree of understanding are presented. For example, the uniform delay equation is developed and used as a means of representing intersection performance. However, the second and third terms of the Highway Capacity Manual delay equation are not included, as students will have no basis for understanding the foundation of these terms.

- Learning objectives are clearly stated at the beginning of each section so that the student knows what is to come. At the end of each section, the learning objectives are reiterated along with a set of concepts that students should understand once they complete the work in the section.

- Over 70 figures are included in the module. We believe that graphically illustrating basic concepts is an important way for students to learn, particularly for queuing model concepts and the development of the change and clearance timing intervals.

- Over 50 computational problems and two field exercises are provided to give students the chance to test their understanding of the material.

We hope that university instructors find this material to be useful in helping them to bring a more active learning and concepts-based approach to their classrooms with the objective of developing a deeper understanding of traffic control systems by their students.

Michael Kyte
Maria Tribelhorn

ACKNOWLEDGEMENTS

We are grateful for funding from the U.S. Department of Transportation's University Transportation Centers Program that supported the work that led to the completion of this module.

We would like to acknowledge the following people for their contributions to this work. University of Idaho graduate student J.J. Peterson conducted some of the research on conceptual understandings of signalized intersections. Marti Ford read a number of drafts and provided detailed and valuable comments on improving this material. Enas Amin provided detailed comments on the text and assisted in the construction of the homework problems and their solutions. Tom Urbanik, Paul Olson, Ed Smaglik, David Hurwitz, Kevin Chang, Anuj Sharma, Zong Tian, and John Zachar provided extremely helpful comments and perspective. University of Idaho civil engineering students used earlier versions of this material and, through their work, showed us where improvements were needed. We are grateful to each of these individuals as well as others who helped in many other ways.

Contents

List of Tables

List of Figures

1. OVERVIEW OF INTERSECTION OPERATION AND CONTROL

> **Learning objectives:**
> - Describe common arterial and intersection problems
> - Describe the urban street system and its components
> - Describe intersection operations and design objectives

1.1 Traveling Along an Urban Street: The Big Picture

What do you expect from the experience of traveling from one place to another on the streets in the city or town in which you live? Sometimes the experience is uneventful. You ride your bike along a path that is safely separated from other traffic. You walk along a well-marked sidewalk and are able to cross a street quickly. When you arrive at an intersection in your car, the display is green and you can continue through the intersection without stopping. The light rail train you are riding on has priority at intersections and is rarely delayed by other traffic. The truck that you drive along the same route each workday can reliably make deliveries to your customers because the travel times are predictable each day.

But too often we hear complaints like the following:

- I have to battle cars at each intersection; there's no safe way to ride my bike on these crowded streets.
- There are no sidewalks on my route to school and the crossing times at major intersections make it difficult for me to safely cross before the red light comes up.
- I have to stop at each intersection that I come to. And sometimes it seems like everyone is stopped in all directions! Who's timing the lights anyway?
- Light rail just isn't all that convenient. We've got to fight traffic just like it was driving my car.
- My customers aren't happy about their daily deliveries. Sometimes I drop off their orders by noon, other days it's later in the afternoon. This unreliability hurts their businesses and mine.

Transportation engineers experience these same kinds of problems in their daily travels. But they see them through a different lens. They bring a wider perspective to these problems by defining them differently and having a framework for classifying and evaluating them. They also have analytical tools to measure the performance of an existing facility or predict the performance of a new one. They have a toolbox of solutions to help them better manage congestion, reduce the potential for crashes, and make the experiences for bicyclists, pedestrians, transit riders, and automobile drivers safer and more reliable. This process of analysis and design, followed by evaluation, allows transportation engineers to address such problems as:

- How do you manage traffic flow along an arterial to minimize the number of stops at each traffic signal, or if stops occur, to make sure that they don't last long?
- What type of control should be used at a given intersection and what will the level of performance be for each type of control considered?
- How do you serve the bicycle rider, the automobile driver, the delivery truck driver, the transit rider, and the pedestrian at an intersection, each with different needs for safety and protection, and each with differing abilities to accelerate and maintain certain travel speeds?
- How do you make travel more reliable so that periodic incidents or recurring congestion don't result in widely varying times for commuters?

In this module, you will learn about some of the factors involved in the design of one type of intersection control, traffic signals. You will also learn how to forecast the performance of one type of signalized intersection, one operating with pretimed control.

1.2 The System and Its Components

The individual intersections that make up a system of urban streets are commonly controlled by traffic signs and signals.

- When traffic volumes are low, such as in a residential area, yield control is often sufficient. Yield signs require drivers to reduce their speed and determine that no other conflicting vehicles or pedestrians are present before they can safely proceed through the intersection.
- Stop sign control requires that drivers come to a complete stop before entering the intersection. All-way stop-control requires drivers on all intersection approaches to stop, while two-way stop-control requires only drivers on the stop-controlled approaches to stop. While the stop sign provides the control, the interaction and decision-making between drivers and pedestrians on the approaches are also important factors in the safe operation of this type of intersection.
- A roundabout is a circular intersection that is controlled by yield signs. The geometric layout of the roundabout forces drivers to slow as they approach the intersection. The gentle angle at which vehicles approach and enter the intersection eliminates the right angle crashes that often occur at standard intersections with 90-degree approaches. Drivers enter the intersection only when they determine that the gap in the traffic already on the roundabout is large enough for them to safely do so.
- Traffic signals provide control by showing green indications only to those movements that can travel through the intersection together at the same time. A time separation, indicated by the yellow and red displays, is provided between each set of these compatible movements.

1. Overview of Intersection Operation and Control

The geometric layout of the intersection includes the number and width of lanes, the movements served in each lane, and the horizontal and vertical alignment of the approaches to the intersection. These geometric elements must be designed such that sufficient capacity is provided (usually determined by the number of lanes on each approach) so that drivers have sufficient space and sight distance to make safe decisions about entering and traveling through the intersection.

While intersections have been historically designed to serve vehicular traffic, primarily automobile drivers, it is more common today for transportation engineers to consider the needs of other users such as pedestrians and transit riders. Transportation engineers have begun to see the environmental and land-use benefits of giving priority at intersections to transit vehicles such as light rail trains. More travelers in transit and fewer traveling alone in their private automobiles mean less fuel consumed and fewer air pollutants. It also means fewer parking lots required in downtown areas. And, more travelers riding their bikes and walking along streets in which non-auto modes are emphasized means healthier citizens. Setting and reflecting on these goals of moving people and not just vehicular traffic is important for the transportation engineer before beginning the design of a street or an intersection. What do you want to accomplish and how will you measure whether you succeeded?

Most signalized intersections today respond to the variations in traffic demand throughout the day using what is called traffic-actuated control. Users, whether driving in vehicles or walking as pedestrians, are detected when they approach or arrive at an intersection. The traffic controller determines whose turn it is to be served at a given time and for how long. This system can be represented by what is called a traffic control process diagram. This model of the traffic control system has four interrelated components and is shown in Figure 1.

- The user arrives at the intersection and is identified by detectors or sensors.
- The detectors transmit this information to the traffic controller.
- The controller, through a series of timing processes and control algorithms, determines which group of users are served at a given time and for how long, and sends this information to the signal display.
- The user responds to the signal display, and the process continues in this recurring loop.

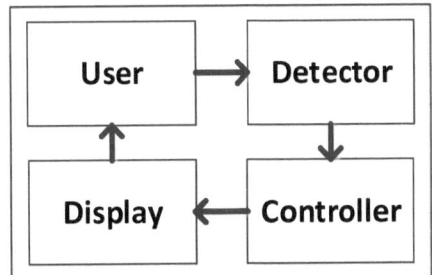

Figure 1. Traffic control process diagram for traffic-actuated Control

1.3 Your Starting Point

A traffic-actuated control system consists of a complex and interrelated set of processes, particularly when each intersection operates as part of a coordinated system. But rather than expect you to learn this complex system as you begin your transition from system user (car driver, pedestrian, bicyclist) to transportation engineer, we will instead focus our work on a more simplified system. This approach will allow you to learn the fundamentals of traffic flow and intersection control before you take on the challenge of a more complex system. Your starting point will be a system with the following characteristics:

- A single intersection operating in isolation apart from any surrounding intersection.
- Traffic flow that arrives at this intersection in a uniform manner at a constant rate of flow.
- Pretimed signal control (represented in the traffic control process diagram shown in Figure 2), in which detectors or sensors are not used, and each group of users is served for a predetermined duration of time.
- Automobile users only.

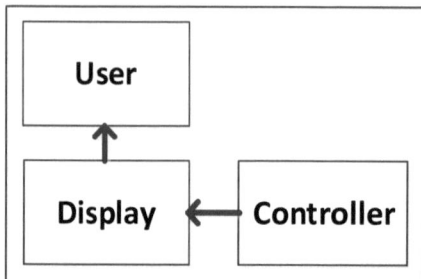

Figure 2. Traffic control process diagram for pretimed control

The culminating activity of this module is the design of a signal phasing and timing plan for an intersection, and the evaluation of how well the intersection will perform under this plan. To build the skills and understandings needed to complete this design, you will learn about traffic flow and signal control processes for a pretimed isolated signalized intersection:

- Representing traffic flow arriving at and departing from a signalized intersection using a deterministic queuing model.
- Resolving conflicts between users so that each traffic stream can safely travel through the intersection.
- Safely changing the right of way from service to one group of users to another.
- Allocating scarce green time between different users according to a priority process based on objectives that have been established so that the available capacity can be efficiently utilized.
- Keeping delay as low as possible by limiting the length of the cycle.
- Judging the performance of the system using measures such as delay and queue length.

1. Overview of Intersection Operation and Control

Figure 3 shows the topics covered in the nine sections that make up this module. Each section opens with a statement of the learning objectives for that section. Content is presented that describes important terms and concepts as well as relevant models that will help you to better understand the terms and concepts. Example problems illustrate the application of the models and concepts. Finally, each section ends with a summary of the key concepts presented in that section.

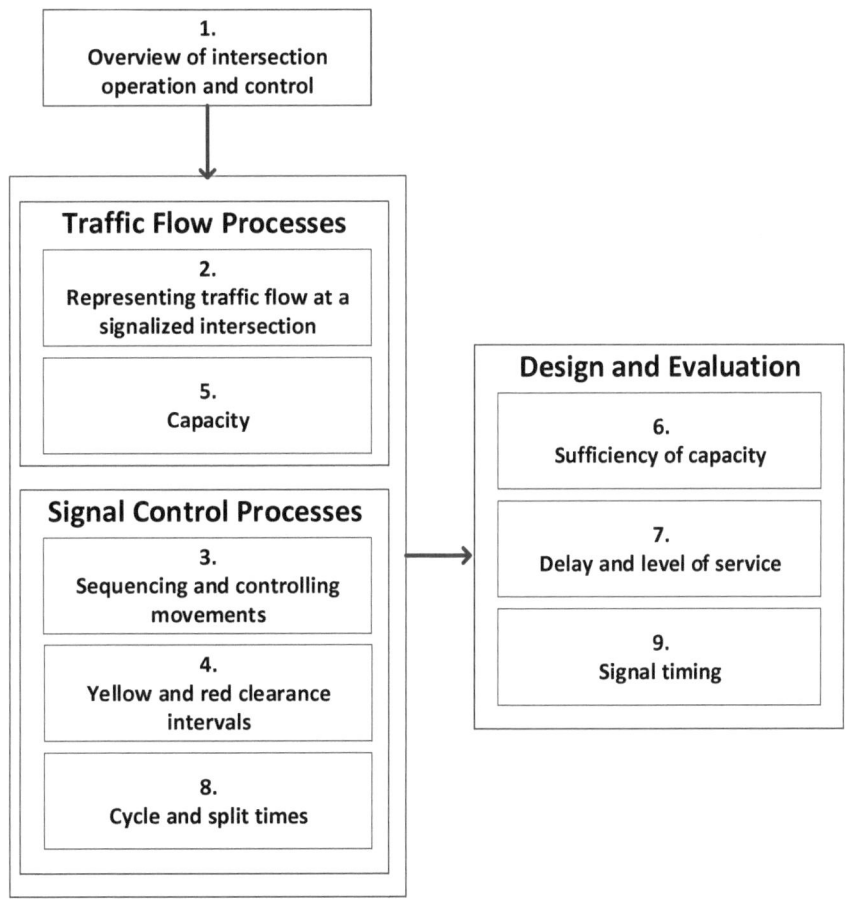

Figure 3. Concepts and topics

Once you master the fundamentals of this more limited system, you can, in future courses, learn about more advanced forms of traffic control that include the design and location of detectors or sensors, determination of the timing processes that respond to varying rates of traffic demand, and the resultant signal displays that indicate which users have right of way and for how long. You will also learn to consider the needs and characteristics of bicycle riders, pedestrians, truck drivers, or bus or light rail vehicle operators. As a look ahead to these more complex systems, we should note three references that provide guidance to transportation engineers on the design and operation of signalized intersections.

- The Manual of Uniform Traffic Control Devices (MUTCD) [1], published by the Federal Highway Administration of the United States Department of Transportation, provides guidance on when a traffic signal is warranted. There are a total of nine warrants which are used to justify the need for a traffic signal. The warrants consider vehicle volumes, pedestrian volumes, school crossings, signal coordination, and crash experience.

- The Traffic Signal Timing Manual (TSTM) [2], also published by the Federal Highway Administration, provides guidance on signal phasing and timing. Most relevant to the material covered in this module, the TSTM covers phasing plans and timing parameter guidance for the green, yellow, and red clearance intervals. The TSTM also covers a number of topics that are beyond the scope of this module, including how signal timing projects are initiated, prioritizing the needs of users of system, the roadway geometry, placement of detectors and how they will function, whether the signal should be coordinated with other nearby signals as a system or operate in isolation, and how to measure intersection performance.

- The Highway Capacity Manual (HCM) [3], published by the Transportation Research Board, provides traffic analysis tools to assist the transportation engineer in determining the quality of service that would result at a signalized intersection given traffic flow rates, intersection geometry, and the traffic signal control plan. The HCM models produce estimates of capacity, delay, and queue length, and link delay to level of service.

2. REPRESENTING TRAFFIC FLOW AT A SIGNALIZED INTERSECTION

> **Learning objectives**:
> - Describe traffic flow characteristics on an intersection approach
> - Apply a time-space diagram to describe flow parameters
> - Describe the operation of a signalized intersection as a queuing process
> - Represent the operation and performance of a signalized intersection using a flow profile diagram, a cumulative vehicle diagram, and a queue accumulation polygon

A model is often used in engineering to represent the behavior of a physical system. The model includes only those elements of the system that are relevant to the problem of interest. A model requires a set of input parameters that specify the state of the system. The model then produces a set of output parameters that help the engineer to evaluate the system performance based on the input state. In this section, a queuing model is used to represent traffic flow at a signalized intersection. Based on the state of the system (as represented by traffic volumes and signal timing parameters), the queuing model produces performance measures such as delay and queue length.

2.1 Vehicle Trajectories at a Signalized Intersection

The flow of vehicles approaching and traveling through a signalized intersection can be represented by a time-space diagram. A time-space diagram shows the position of each vehicle at any point in time, and the slopes of its trajectories show the speed of a vehicle at any point in time. The vehicular signal display intervals of green and red are also shown. The time that it takes to cycle through the display of these intervals is called the *cycle length*.

The time-space diagram in Figure 4 shows the flow of individual vehicles traveling through a signalized intersection (blue lines from lower left to upper right in the figure) during two complete signal cycles, illustrating several important concepts.

- During the first cycle, three vehicles arrive at the intersection during the green indication and travel through the intersection without stopping. The *arrival headway* h_a (the headway[1] between vehicles arriving at a signalized intersection) is constant, a flow pattern called *uniform flow*. We will assume the condition of uniform arrival flow in the queuing models that will be considered shortly.
- During the second cycle, three vehicles arrive at the intersection during the red indication. A fourth vehicle arrives during the green indication but must initially stop because the vehicle in front of it is stopped. As each of these four vehicles arrives, a queue (waiting line) forms at the stop line. The trajectory of the four vehicles is horizontal during red, illustrating that while time passes, the vehicle positions are stationary. The position of each vehicle

---

[1] *Headway* is defined as the time between the passage of successive vehicles at a given point.

in the queue, and the spacing between vehicles in the queue, is shown in the time space diagram.

- At the beginning of the green indication during cycle 2, the vehicles begin to enter the intersection. The headway between vehicles in the departing queue is called the *saturation headway* h_s.
- A final vehicle arrives during cycle 2 after the queue has cleared and travels through the intersection without stopping.

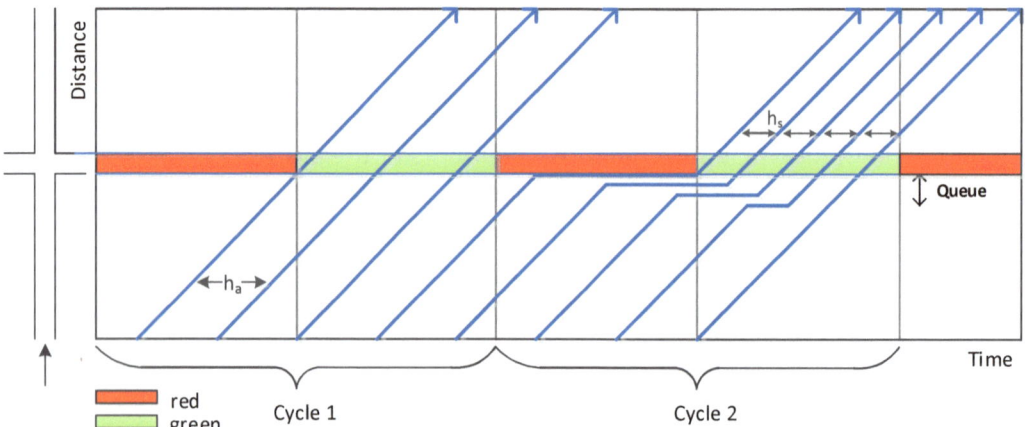

Figure 4. Vehicle trajectories (blue lines) represented in a time-space diagram

2.2 The Queuing Process at a Signalized Intersection

Traffic flow at a signalized intersection can be represented by a standard queuing model, known as the *D/D/1 model*. The D/D/1 model assumes a deterministic arrival pattern, a deterministic service pattern, and one service channel. A deterministic arrival or service pattern means that the pattern is known and does not vary over time. Stated another way, there is no randomness in the pattern. One service channel implies one lane of an intersection approach. The model also assumes that the demand is less than the capacity.

Figure 5 shows these elements as applied to one lane on an approach to a signalized intersection. The *server* is the first vehicle position at the stop bar. This is the point at which vehicles are served as they exit the queuing system and enter the intersection. The *queue* forms behind the vehicle in the server and extends to some maximum point, depending on the arrival and service rates of the system. While queuing theory assumes that this line of vehicles is a vertical stack, in reality the queue extends horizontally upstream from the stop bar to the point of the maximum queue. This point is considered to be the entry point to the queuing system.

2. Representing Traffic Flow at a Signalized Intersection

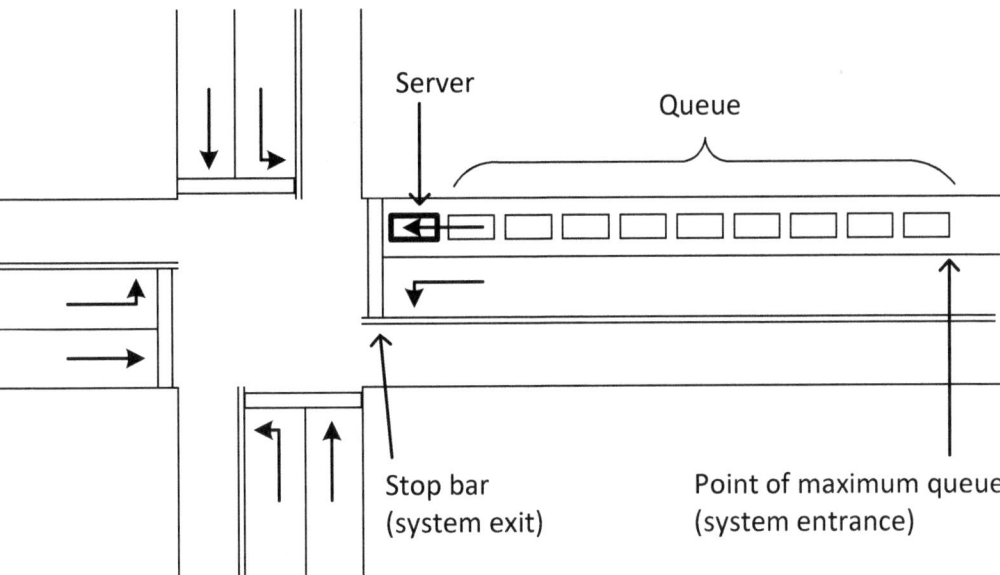

Figure 5. Elements of queuing system

Figure 6 shows another view of the queuing process, here overlaid vertically on a time-space diagram (at the left side of the figure). The arrival pattern is shown just upstream of the signalized intersection, showing the constant headway between each of the vehicles. As noted above, this pattern is called *uniform arrivals* and is consistent with the deterministic arrival pattern for the D/D/1 queuing model. The service or departure pattern is shown just downstream of the intersection for three time periods.

- During period 1, vehicles 1, 2, and 3 arrive during the red interval and form a queue. The service or departure rate is zero because the signal indication is red.

- When the signal display changes to green, vehicles 1, 2, and 3 depart from the intersection at the saturation flow rate. Here the headway between vehicles is equal to the saturation headway. A fourth vehicle arrives during green but joins the queue and can't be served until the queue clears. The service rate for this vehicle is also equal to the saturation flow rate. The time that it takes for the queue to clear (called the *queue service time* g_s) is the duration of the second period.

- During the third period, vehicles 5 and 6 arrive and depart at a constant rate, equal to the arrival rate. Since the queue has cleared, these vehicles experience no delay.

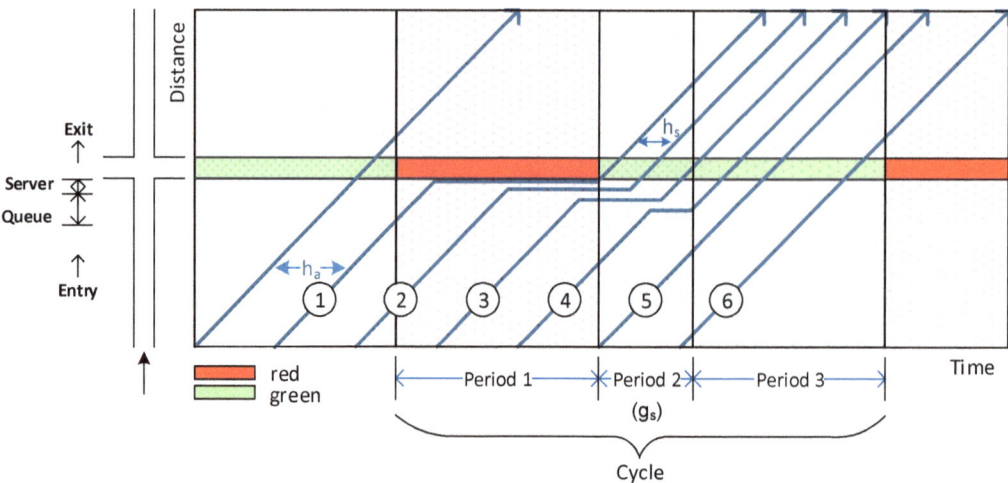

Figure 6. Vehicle trajectories arriving at and departing from a signalized intersection

There are three other ways to represent the queuing process including the flow profile diagram, the cumulative vehicle diagram, and the queue accumulation polygon. These diagrams also show important concepts in the traffic flow process such as capacity and delay, as well as other ways to represent intersection operation and performance, and are discussed below.

2.3 The Flow Profile Diagram

The *flow profile diagram* (Figure 7) represents both the arrival flow to and the departure flow from a signalized intersection over time. The flow profiles can be extracted from the time-space diagram shown in Figure 6 either by calculating the flow rate from the headway between vehicles, or by counting the number of vehicles over constant time segments. The arrival flow rate is constant (uniform) and is represented by the solid line in Figure 7. The service flow rate (represented by the dashed line) varies during the cycle, according to the three time periods noted in the discussion of the time-space diagram above. The service flow rate is equal to:

- Zero, during the red indication.
- The saturation flow rate s, while the queue is clearing, during an interval called the queue service time g_s.
- The arrival flow rate v, after the queue clears and until the end of the green interval.

Since the queue clears before the end of green, the volume is less than the capacity. This meets one of the assumptions of the D/D/1 queuing model established earlier.

This service pattern is repeated for each signal cycle.

2. Representing Traffic Flow at a Signalized Intersection

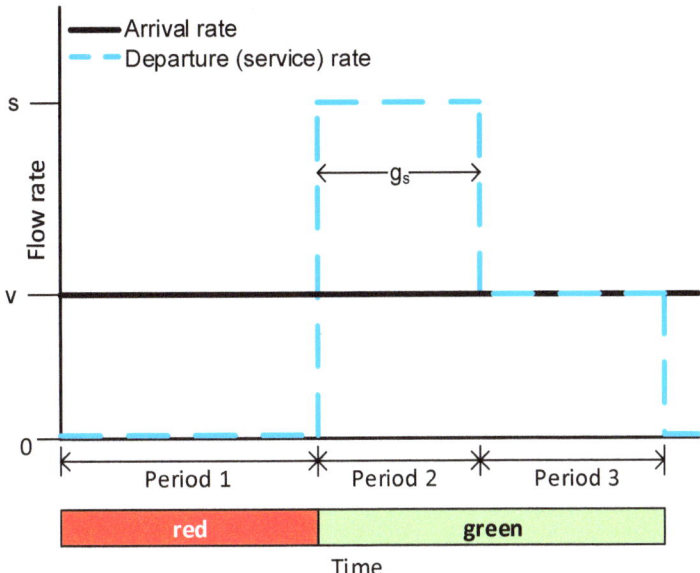

Figure 7. Flow profile diagram

Figure 8 shows that the server can be represented by a pipe with the arrival (input) flow profile diagram shown on the left and the service (output) flow profile shown on the right. The diameter of the pipe approximates the saturation flow rate.

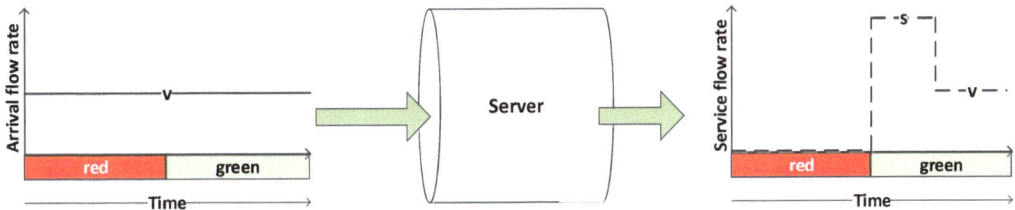

Figure 8. Queuing process showing input, server, and output

Example 1. Flow Profile Diagram
Prepare a flow profile diagram that represents the following conditions:
- Arrival flow rate = 600 veh/hr
- Saturation flow rate = 1900 veh/hr
- Queue service time = 13.8 sec
- Cycle length = 60 sec
- Green time = 30 sec
- Red time = 30 sec

The arrival flow profile is represented by a horizontal line whose value is a constant 600 veh/hr. The departure or service flow profile is represented by three line segments, one for each of the three periods described above:
- During red, the departure flow rate is zero.
- During the period that the queue is clearing, or the queue service time, the departure flow rate is equal to the saturation flow rate, or 1900 veh/hr.

11

- After the queue has cleared, the departure rate is equal to the arrival rate, or 600 veh/hr.

Figure 9 shows both the arrival and departure flow profiles for the conditions just described.

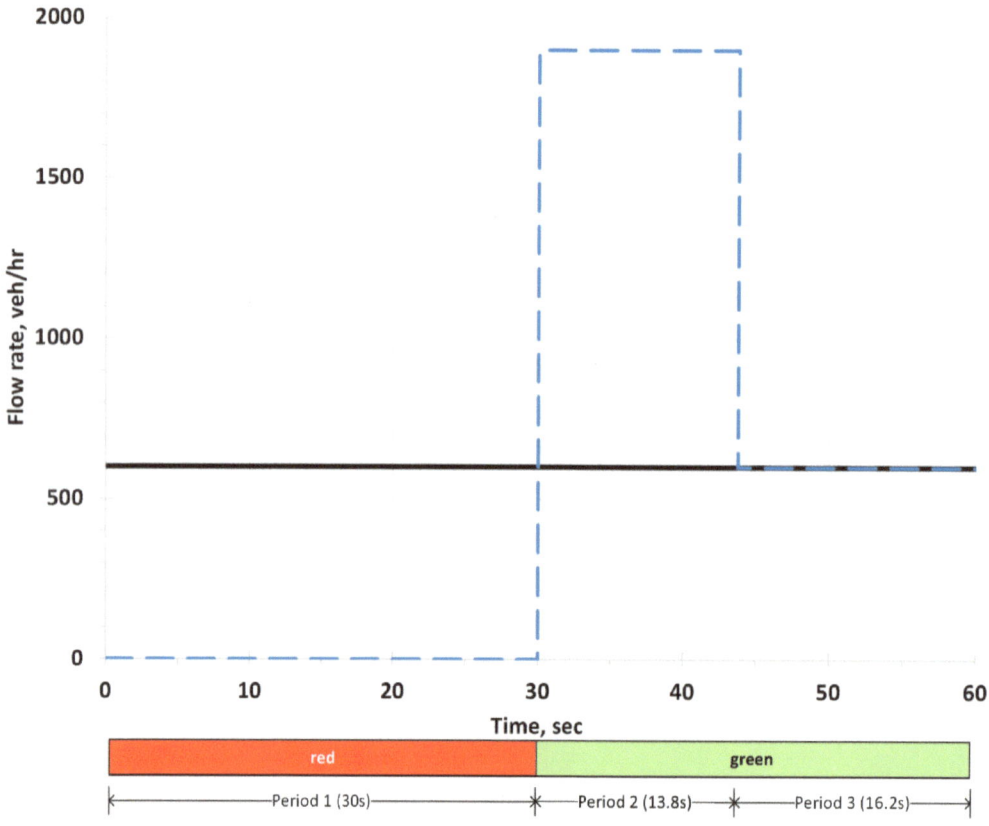

Figure 9. Example flow profile diagram

2.4 The Cumulative Vehicle Diagram

The *cumulative vehicle diagram* (Figure 10) is a running total of the number of vehicles that have arrived at and departed from the intersection over time. The cumulative vehicle diagram shows two lines, one representing the cumulative number of arrivals over time and the other the cumulative number of departures.

When we assume that the arrival pattern is uniform, with constant headways, the horizontal line from the flow profile diagram becomes a line of constant slope in the cumulative vehicle diagram. The slope of the line representing vehicle arrivals is equal to the arrival flow rate. As we move from left to right in the diagram (representing the passage of time), the y-axis value shows the running total of the number of vehicles that have arrived at the intersection at any point in time.

The service pattern is again divided into three periods. During the red indication, no vehicles can depart from the intersection, so the total number of departures is zero during this period. Once the green is displayed, vehicles begin to depart from the intersection at the saturation flow rate, with headways between vehicles equal to the saturation headway. The slope of the departure line during this time interval is equal to the saturation flow rate. At the point that the queue clears, the arrival line and the departure line become coincident. Their slopes are the same and equal to the vehicle arrival rate during the remainder of the green interval.

The cumulative vehicle diagram also shows three measures of intersection performance, indicating how well the intersection is operating.

- The first measure is the length of the queue at any point in time. The *length of the queue* is the difference between the number of vehicles that have arrived at and departed from the intersection at any point in time. Graphically, the queue length (or number of vehicles currently in the system, Q) is the vertical distance between the arrival line and the departure line, as illustrated in Figure 10.

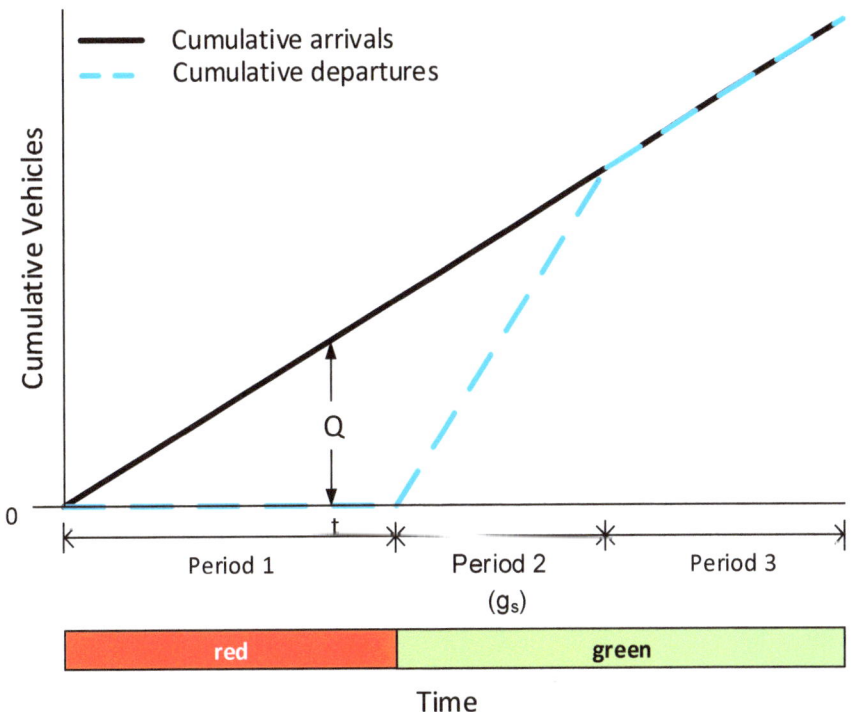

Figure 10. Cumulative vehicle diagram showing queue length at time t

- The second measure is the time that each vehicle spends in the system, or the *delay* that it experiences. Consider the horizontal line connecting the arrival line and the departure line for the vehicle (noted as vehicle i) shown in Figure 11. Point 1 on the arrival line is the time that the vehicle enters the system; point 2 on the departure line is the time that the vehicle exits the system. The time interval between these two points is the delay experienced by the vehicle.

13

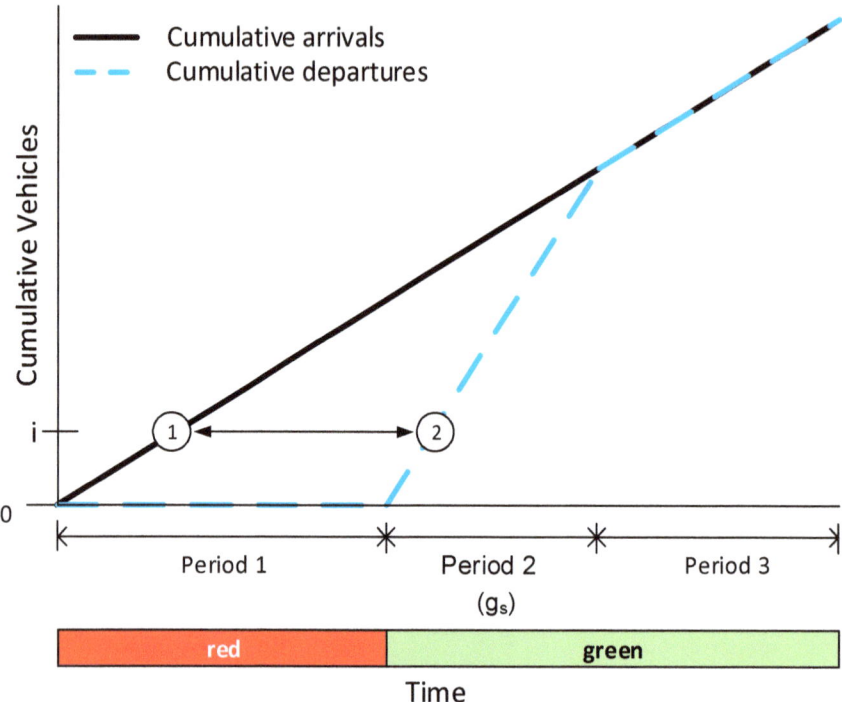

Figure 11. Cumulative vehicle diagram showing delay for vehicle i

- The third measure is the *total delay* experienced by all vehicles that arrive at and travel through the intersection. If we add all of the horizontal lines described in the bullet above for all vehicles, we get the total delay experienced by all vehicles. The total delay is the area of the triangle formed by the arrival and departure lines and shown as the shaded area in Figure 12.

2. Representing Traffic Flow at a Signalized Intersection

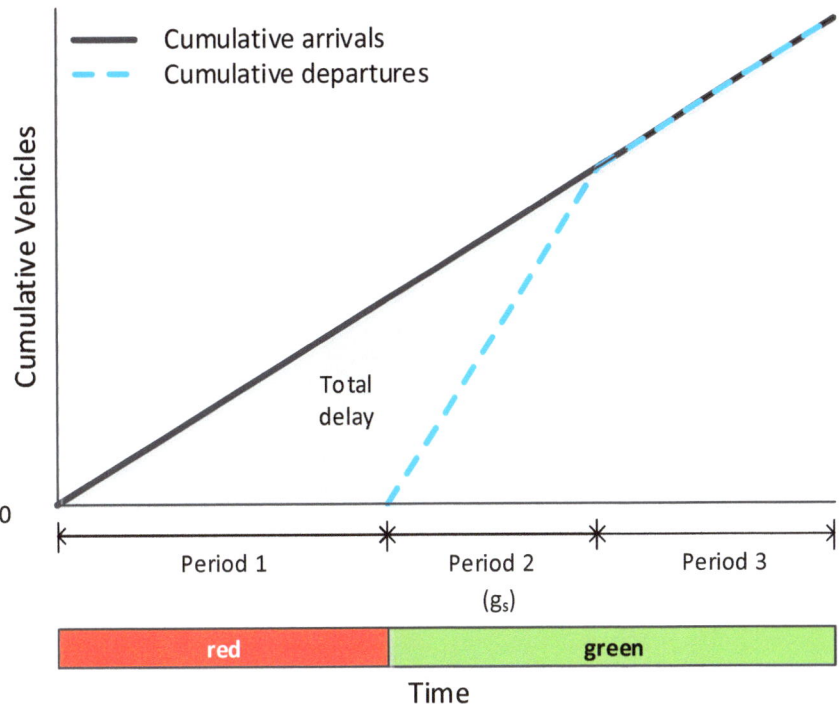

Figure 12. Cumulative vehicle diagram showing total delay

Example 2. Cumulative Vehicle Diagram
Field data collected on one approach of a signalized intersection showed that:
- Vehicles arrive every 6 sec at a uniform rate.
- The cycle length is 60 sec, with red and green time intervals of 30 sec each.
- Vehicles depart every 2 sec after the beginning of green.
- The queue service time is 14 sec.

Prepare a cumulative vehicle diagram that represents these conditions.

The cumulative vehicle diagram for these conditions is shown in Figure 13. Since a vehicle arrives at the intersection every six seconds, a cumulative total of five vehicles arrive from the beginning of red until the end of red. We've assumed that the vehicle that arrives at t = 0 (end of green) travels through the intersection without stopping. Also, since we are dealing with the discrete events of vehicle arrivals in the field, the lines are "stair step" with each increase representing a vehicle arrival or departure. By contrast, the theoretical depictions presented earlier are based on a continuous and not discrete process.

The chart shows that at t = 44 sec, the cumulative number of vehicles that have entered the system (seven) equals the number that have exited. It is at this point (14 sec after the start of the green interval) that the arrival and departure lines become coincident and the queue clears.

15

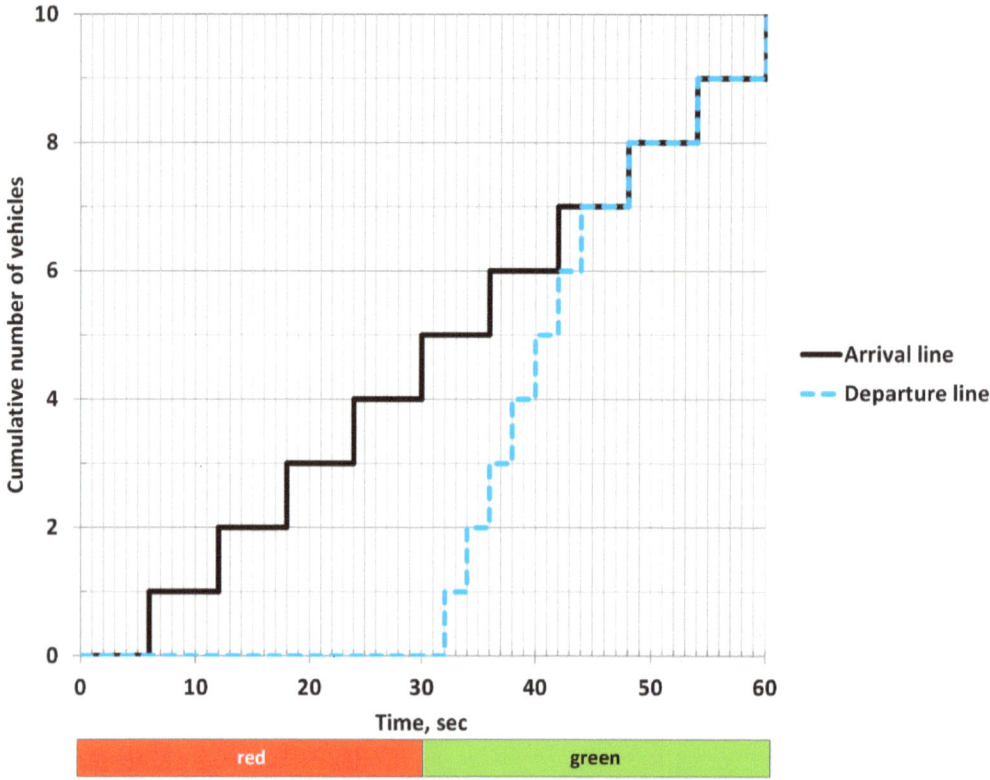

Figure 13. Example cumulative vehicle diagram

What is the delay for the vehicles that arrive at the intersection? Table 1 shows the times that each vehicle arrives at and departs from the intersection, as read from the cumulative vehicle diagram (Figure 13) for the first seven vehicles. The difference between the arrival and departure times is the time in the system or the delay experienced by each vehicle.

Table 1. Vehicle arrivals and departures (and time in system)

Vehicle number	Arrival time (sec)	Departure time (sec)	Delay (time in system), (sec/veh)
1	6	32	26
2	12	34	22
3	18	36	18
4	24	38	14
5	30	40	10
6	36	42	6
7	42	42	2
		Total	98 sec

Vehicle #1 has the longest delay (26 sec) since it arrives near the beginning of the red interval. The seventh vehicle arrives just as the queue is clearing and has a delay of about 2 sec. Vehicles 8, 9, and 10 arrive and leave without delay, as the queue has cleared by the time that vehicle 8 arrives. The total delay for all vehicles is 98 sec.

2.5 The Queue Accumulation Polygon

The *queue accumulation polygon* represents the length of the queue at any point in time and is derived from the cumulative vehicle diagram. Figure 14 shows the queue accumulation polygon, again for the case of uniform arrivals: the queue grows during red and reaches its maximum length at the end of the red interval (or the beginning of the green interval). It decreases once the green interval begins and reaches zero when the queue clears. It remains at zero until the end of the green interval.

The area of the queue accumulation polygon is equal to the area between the arrival and departure lines of the cumulative vehicle diagram: both areas represent the total delay experienced by all vehicles that arrive and leave during the cycle.

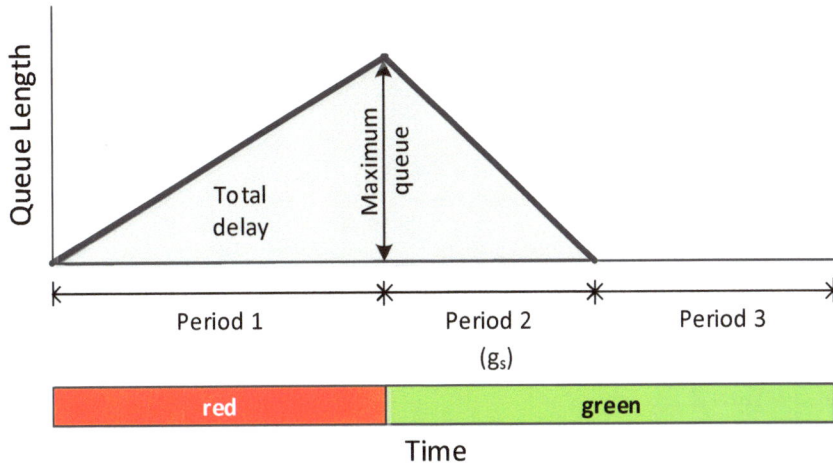

Figure 14. Queue accumulation polygon

Example 3. Queue Accumulation Polygon

Consider the conditions given for the cumulative vehicle diagram in Example 2. Prepare a queue accumulation polygon that represents these conditions.

The queue accumulation polygon represents the length of the queue over time and is shown in Figure 15 for the given conditions. The queue accumulation polygon shows the queue growing during red and reaching a maximum length of 5 vehicles at the end of red (t = 30 sec). The queue begins to clear when the display changes from red to green and clears 14 sec later (t = 44 sec). The queue remains at zero after t = 44 sec, as vehicles arrive and depart without delay. Figure 15 represents the discrete form of the queue accumulation polygon where the change for each vehicle is shown, in contrast to the continuous form shown in Figure 14.

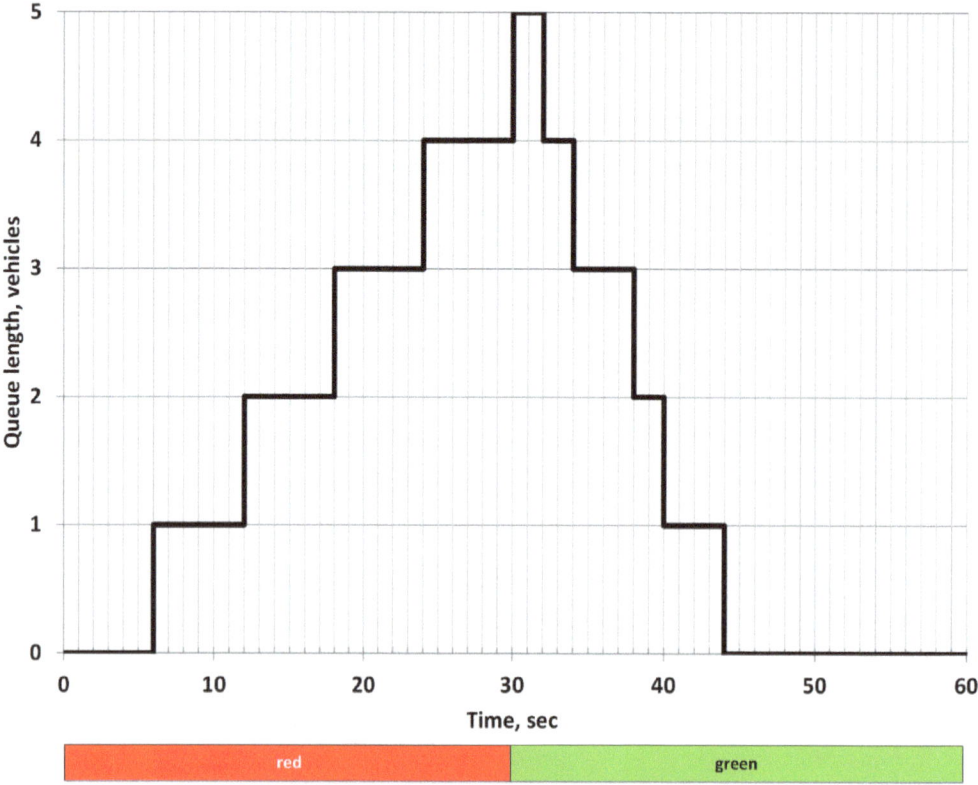

Figure 15. Example queue accumulation polygon

2.6 Summary of Section 2

What You Should Know and Be Able To Do:

- Describe traffic flow characteristics on an intersection approach
- Apply a time-space diagram to describe flow parameters
- Describe the operation of a signalized intersection as a queuing process
- Represent the operation and performance of a signalized intersection using a flow profile diagram, a cumulative vehicle diagram, and a queue accumulation polygon

Concepts You Should Understand:

- Concept 2.1: Queuing process at signalized intersection

We have represented traffic flow on one lane at a signalized intersection as a queuing process. The arrival pattern consists of uniform flow with a constant rate. The service pattern is represented by three values: zero during red, the saturation flow rate during the queue clearance process, and the arrival flow rate after the queue has cleared. The process can be represented by a flow profile diagram.

- Concept 2.2: Queuing process representation

The queuing process can also be graphically represented as a cumulative vehicle diagram and a queue accumulation polygon. Each diagram shows the evolution of the queue length during the cycle (the growth during red and the clearing

2. Representing Traffic Flow at a Signalized Intersection

during green) and the total delay experienced by all users of the system (the areas of the triangles).

3. SEQUENCING AND CONTROLLING MOVEMENTS

Learning objectives:
- Define and apply the terms movement and phase
- Describe the sequencing and control of movements at a signalized intersection
- Determine left turn treatment
- Draw and interpret a ring barrier diagram that represents a particular phasing plan
- Define and apply the term stage

In the previous section, we considered traffic flow on one approach of a signalized intersection. In this section, we consider traffic flow at the entire intersection and how the movements on the individual approaches are sequenced and controlled. Safety and efficiency are the two primary goals of intersection control. Considering efficiency, as many movements as possible should be served concurrently. But for safety reasons, conflicting movements must be served during different periods of the signal cycle, separated in time by the yellow and red clearance intervals.

3.1 Movements

A standard intersection with four approaches can have up to twelve vehicular movements. These movements, and the way in which they are typically numbered and referred to, are shown in Figure 16. The numbering scheme is based on the National Electrical Manufacturers Association (NEMA) standard.

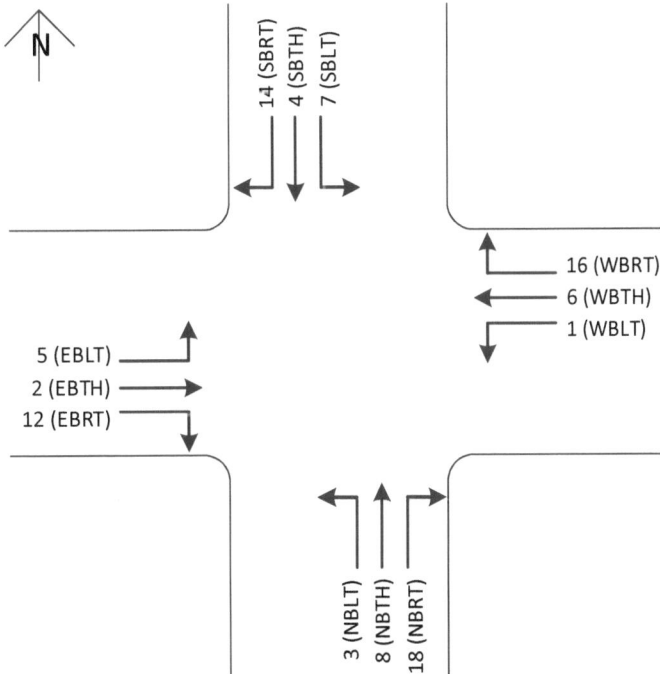

Figure 16. Numbering and notation of movements at a signalized intersection

A *movement* is defined by the direction of travel from its origin and the turning maneuver that a vehicle completes to its destination. For example, movement 2 begins traveling in the eastbound direction and continues through the intersection in the eastbound direction. It is referred to as an eastbound through movement (abbreviated as EBTH). Or, movement 1 begins traveling in the westbound direction, making a left turn and continuing in the southbound direction. Movement 1 is referred to as a westbound left turn (WBLT) movement. As a final example, movement 14 is referred to as a southbound right turn (SBRT) movement.

A movement is also categorized by any restriction that may be placed on it. There are four such categories:

- An unopposed movement is just that: there is no other movement that opposes this movement. For example, a movement on a one-way street is unopposed.

- A protected movement may have a movement that can oppose it but the signal indication gives the protected movement the right-of-way. For example, a left turn movement may be protected if the signal indication is a green arrow while the opposing traffic movement has a red indication.

- A permitted movement is allowed to travel through the intersection, but must yield if a higher priority opposing movement is present. For example, a permitted left turn may enter and travel through the intersection as long as there are no opposing through movements also desiring to travel through the intersection at the same time.

- A movement can also be prohibited, or not allowed. This restriction can be complete or in effect only during certain periods of the day. For example, left turns can be prohibited (not allowed) during peak periods, especially if the left turn movement shares a lane with a through movement.

Groups of movements are also classified as either *compatible* or *conflicting*. In general, north-south movements conflict with east-west movements. North-south movements are part of a group called a *concurrency group* since these movements may travel concurrently; similarly east-west movements are part of the east-west concurrency group. The concept of the concurrency group is illustrated in Figure 17. Depending on the signal phasing plan (*phase* is defined in the next subsection) and restrictions on the movements, a movement in the north-south concurrency group may be served at the same time as any other movement in this group. The same concept applies to the movements in the east-west concurrency group.

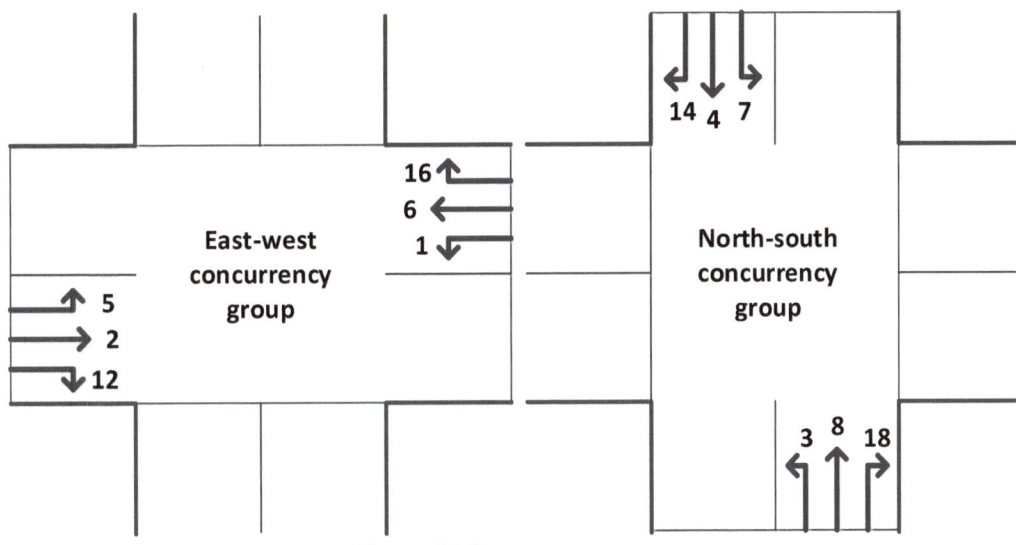

Figure 17. Concurrency groups

3.2 Phasing and the Ring Barrier Diagram

While stop and yield controlled intersections require judgment before a driver can safely enter the intersection, signal control gives an unambiguous indication whether a particular movement has the right of way or not. This right of way assignment is done through the signal display of a green ball, a green arrow, or a flashing yellow arrow.

A *phase* is a timing unit that controls one or more compatible movements at a signalized intersection. The timing unit consists of the consecutive displays of the green, yellow, and red indications shown to the movements controlled by the phase, as shown in Figure 18.

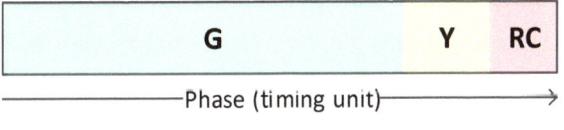

Figure 18. Phase

A *ring* is a sequence of phases that must be served one after the other. Phases are sequenced to separate conflicting movements and to make sure that those movements that are served concurrently are compatible.

A *ring barrier diagram* is the tool that is used to define those movements that are compatible and can be served concurrently, and those that conflict and must be served in sequence. The ring barrier diagram is built upon the concept of the concurrency group described above. The movements of the east-west concurrency group are served first, followed by service to the north-south concurrency group.

An example ring barrier diagram with eight phases and a two ring structure is shown in Figure 19. Barriers separate the two concurrency groups. The phase number is shown in the upper left corner of each square, and the movements that are controlled by that phase are shown with an arrow and movement number. A dashed line indicates permitted movements.

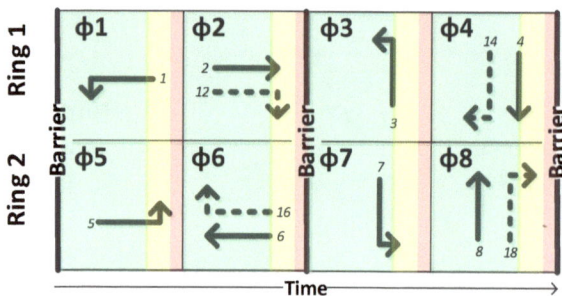

Figure 19. Ring barrier diagram for two ring, eight phase operation

The following rules apply to a ring barrier diagram:

- Each ring shows the order in which phases are sequenced.
- A phase in one ring can time concurrently with a phase in another ring as long as both phases are in the same concurrency group.
- A phase in one concurrency group cannot time concurrently with a phase in another concurrency group. (The exception to this rule is the concept of the *overlap*, which is beyond the scope of this module.)
- A barrier separates the north-south and the east-west concurrency groups. Both rings must cross the barrier at the same time.
- After a phase is served, a change and clearance interval (yellow indication and red indication) provides a time separation between that phase and the next conflicting phase.

3.3 Left Turn Phasing

There are several ways in which left turn movements can be served at a signalized intersection. Figure 19 showed the case of *leading left turns* (served before the through movements) that are *protected* (with no opposing through movements served at the same time). Protected left turns (shown with solid lines) are generally used when the combination of the left turn movement volume and its opposing through movement volume is high.

When left turn volumes are low, *permitted left turn* operation (shown with dashed lines) is possible. This results in a single ring two phase operation, as shown in Figure 20. Here, the left turn and through movements within each concurrency group are controlled by the same phase.

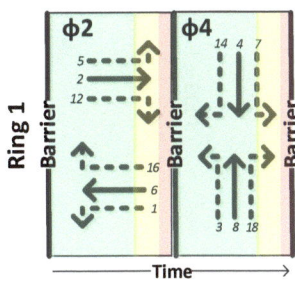

Figure 20. Two-phase ring barrier diagram

Another phase structure is called *split phasing*, in which each approach is served in sequence. Split phasing is used when safety or geometric restrictions don't allow opposing left turns to be served at the same time. This structure can be represented by a single ring, as shown in Figure 21.

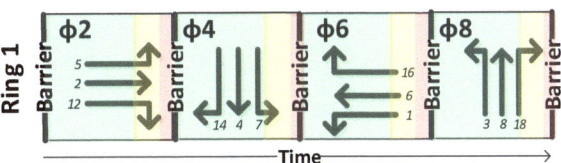

Figure 21. Split-phase ring barrier diagram

The decision whether to provide protected left turn phasing is based on the combination of the left turn and opposing through traffic volumes, the geometric layout of the intersection, the speeds of the opposing traffic, the number and kinds of traffic crashes that have occurred, and the delay and degree of queuing experienced by left turn traffic.

One common guideline is the "cross product" of the left turn volume and the sum of the opposing through and right turning volumes. The Highway Capacity Manual offers the following criterion for this guideline: the use of a protected left turn phase should be considered when, during the peak hour, the product of the left turning volume and the opposing traffic volume equals or exceeds:

- 50,000 if there is one opposing lane,
- 90,000 for two opposing lanes, and
- 110,000 for three or more opposing lanes.

Example 4. Left Turn Phasing
Consider the intersection shown in Figure 22, with one or two through lanes and an exclusive left turn lane on the four approaches. The hourly flow rates for each movement are also shown in the figure.

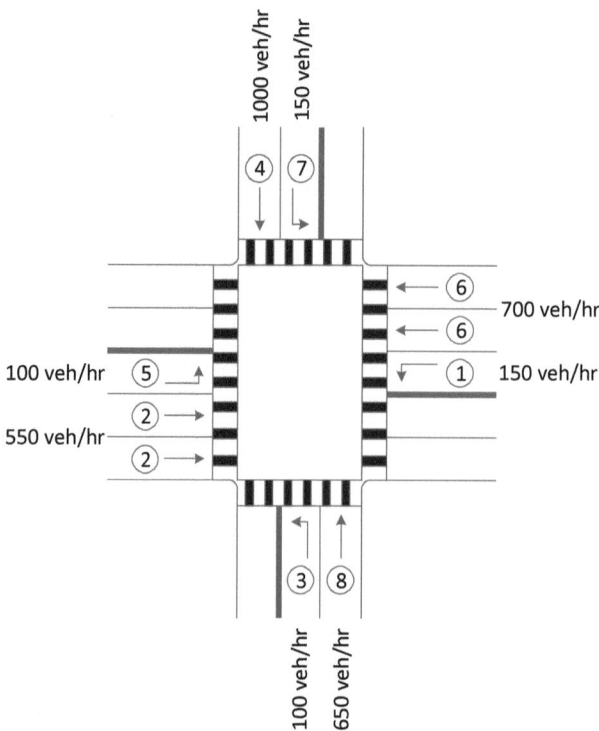

Figure 22. Flow rates for Example 4

Based on these flow rates and the intersection geometry, what left turn phasing would you recommend for each of the four approaches?

The flow rates for the left turn and through movement combinations are shown in Table 2. The cross products for each left turn-through movement combination are computed and also shown in the table. Finally, based on the criteria from the Highway Capacity Manual listed above, the recommended left turn phasing is given. In this case, the cross products are high enough for the north-south movements that protected left turn phasing would be recommended. However, permitted left turn phasing is sufficient for the east-west movements.

Table 2. LT-TH movement cross products and left turn phasing

Movement	Flow rate (veh/hr)	Cross product	HCM guideline	Recommended LT phasing
NBLT	100	100,000	50,000	Protected
SBTH	1000			
SBLT	150	97,500	50,000	Protected
NBTH	650			
EBLT	100	70,000	90,000	Permitted
WBTH	700			
WBLT	150	82,500	90,000	Permitted
EBTH	550			

Example 5. Creating a Ring Barrier Diagram

A signalized intersection has lagging protected left turns for the NB and SB movements and permitted left turns for the EB and WB movements. Create a ring barrier diagram for this case.

Figure 23 shows the ring barrier diagram that represents the conditions described above. For the east-west concurrency group, one phase (in this case phase 2), controls all of the movements. For the north-south concurrency group, phases 4 and 8 (controlling the through and right turn movements) "lead" the left turns (controlled by phases 3 and 7). The left turns "lag" the through movements.

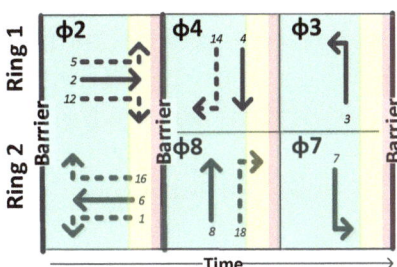

Figure 23. Ring barrier diagram for conditions in Example 5

3.4 Timing Stages

The ring-barrier concept allows compatible phases in different rings (within each concurrency group) to operate for different time durations based on the level of traffic volume. For example, if the traffic volume for the EBLT movement is greater than the volume for the WBLT movement, phase 5 (controlling the EBLT movement) can provide a longer green duration (or "time longer") than phase 1. Similarly, if the traffic volume for the NBLT movement is greater than the volume for the SBLT movement, then phase 3 can time longer than phase 7.

The conditions described above, represented in Figure 24 through the concept of the timing stage, show the intrinsic efficiency of the ring barrier process. A *timing stage* is an interval of time during which no signal displays change. The horizontal length of the phase is its relative time duration. Time moves from left to right.

- Stage 1 includes the concurrent timing of phases 1 and 5 serving the EBLT and WBLT movements. Because the volume for movement 1 is less than the volume for movement 5, phase 1 terminates before phase 5.
- The second stage is the concurrent timing of phases 2 and 5.
- When phase 5 terminates, phases 2 and 6 time concurrently in stage 3.
- The same process applies to the north-south concurrency group, shown in stages 4, 5, and 6.

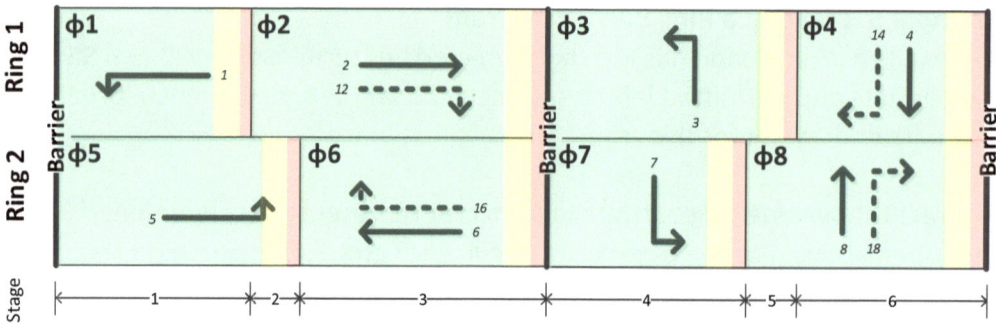

Figure 24. Ring barrier diagram and timing stages

Example 6. Phasing and Timing Stages

Consider the following timing requirements for the eight phases that serve a standard four leg intersection. If protected left turns are required, construct a ring barrier diagram for this intersection that shows the resulting timing stages. Assume also that the left turns lead the through movements.

Table 3. Phase durations

Phase	1	2	3	4	5	6	7	8
Duration, sec	15	30	10	25	10	35	10	25

The ring barrier diagram for leading protected left turns is shown in Figure 25.

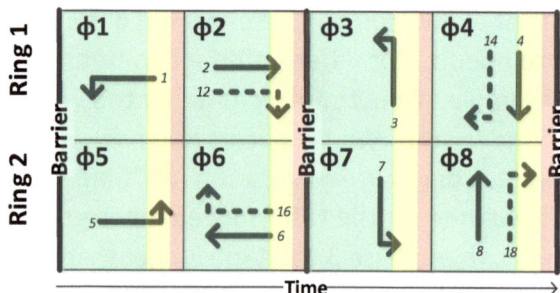

Figure 25. Ring barrier diagram

The ring barrier diagram can also be scaled based on the time required to serve each of the phases noted by the resulting timing stages (see Figure 26).

- For the east-west concurrency group, phases 1 and 5 time concurrently for 10 sec during stage 1
- But phase 1 times for an additional 5 sec. It times concurrently with phase 6 during stage 2.
- During stage 3, phases 2 and 6 time concurrently for 30 sec.
- When the barrier is crossed, phases 3 and 7 time concurrently for 10 sec, as part of stage 4. Here both phases end at the same time.
- Stage 5 consists of phases 4 and 8 timing concurrently for 25 sec.
- Note that the sum of the durations for phases 1 through 4 is equal to the durations for phases 5 through 8.

3. Sequencing and Controlling Movements

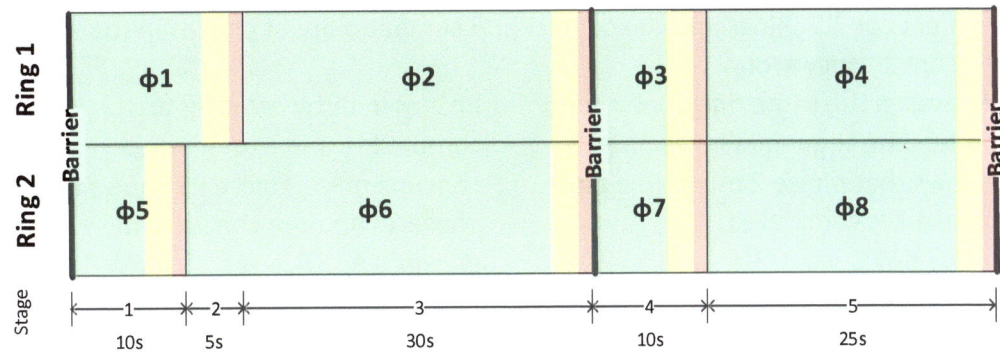

Figure 26. Ring barrier diagram with time scale

3.5 Summary of Section 3

What You Should Know and Be Able to Do:

- Define and apply the terms movement and phase
- Describe the sequencing and control of movements at a signalized intersection
- Determine left turn treatment
- Draw and interpret a ring barrier diagram that represents a particular phasing plan
- Define and apply the term timing stage

Concepts You Should Understand:

- Concept 3.1: Concurrency groups

Figure 27 shows the movements in the north-south and east-west concurrency groups. Movements in one concurrency group conflict with movements in the other concurrency group.

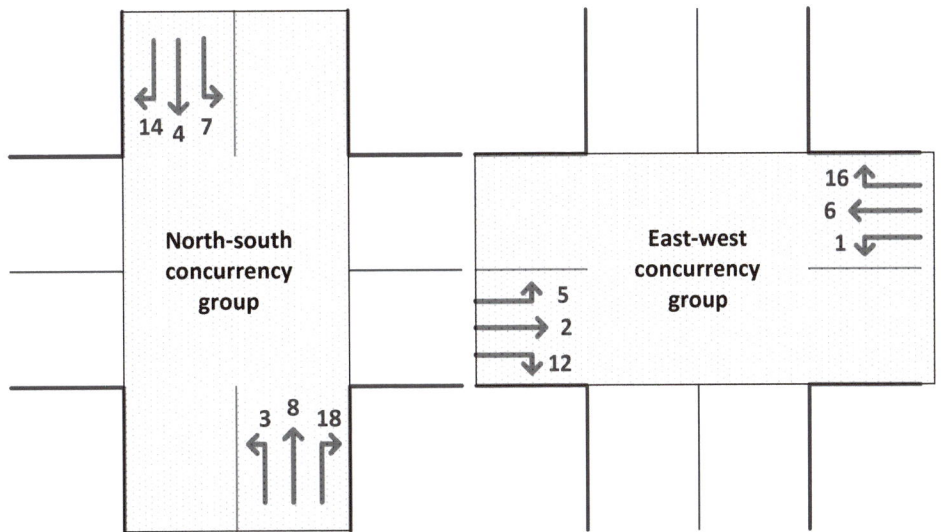

Figure 27. Concurrency groups

- Concept 3.2: Rings and the conflict and compatibility of phases in the same concurrency group:

Phases in the same ring conflict and must time sequentially. Figure 28 shows that phase 2 must time after phase 1 is completed.

Phases in different rings are compatible and can time concurrently. Figure 29 shows that phase 1 can time concurrently with phases 5 and 6.

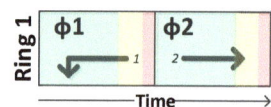

Figure 28. Phases timing sequentially

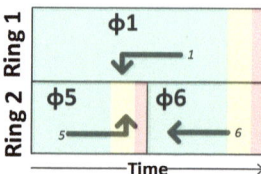

Figure 29. Phases timing concurrently

- Concept 3.3: Ring barrier diagram

The ring barrier diagram shows the phases in each ring that must time sequentially and the barrier that must be crossed at the same time by both rings. An example of a ring barrier diagram for leading protected left turns is shown in Figure 30:

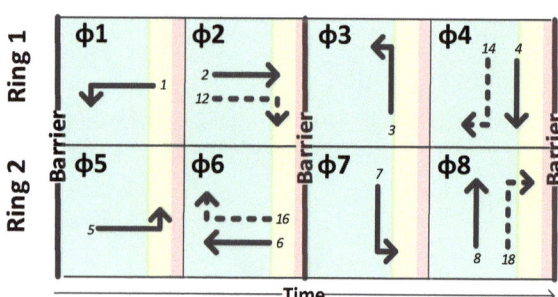

Figure 30. Ring barrier diagram

4. YELLOW AND RED CLEARANCE INTERVALS

Learning objectives:
- Describe the basis for determining the change and clearance intervals
- Calculate the duration of the yellow and red clearance times

The change and clearance timing intervals provide a safety interval between the service to one set of users and another. The duration of these timing intervals must be long enough so that a driver on the intersection approach is able to do one of two things:
1. Safely come to a complete stop at the stop bar, or
2. Clear the intersection before the conflicting traffic begins to enter the intersection.

If the driver is not able to do either of these two things, an unsafe condition is created. An area on the intersection approach in which a driver can neither safely stop nor safely clear the intersection is called a *dilemma zone*. This section provides a model for the calculation of the change and clearance timing intervals that avoids the dilemma zone.

4.1 The Choice Point

At signalized intersections, a safe transition between conflicting phases is provided by the yellow change and red clearance intervals. To calculate the time necessary for the change and clearance intervals, two questions must be considered:
- How long does it take a driver to perceive the need to stop and then brake to a stop?
- How long does it take a driver to safely and completely clear the intersection?

To answer these questions, we will consider the concept of the choice point. Figure 31 shows an arterial in which a vehicle travels from the bottom of the figure to the top, and a time-space diagram where vehicle trajectories can be shown. The near and far sides of the intersection are shown as the horizontal lines in the time-space diagram.

The *choice point*, shown in Figure 31, is the closest point upstream of the stop bar, at the onset of yellow, at which the driver will be able to safely stop should he or she choose to do so. If the driver is any closer to the intersection than this choice point, he or she would not be able to safely come to a stop. The distance from the choice point to the stop bar is equal to the *stopping distance*.

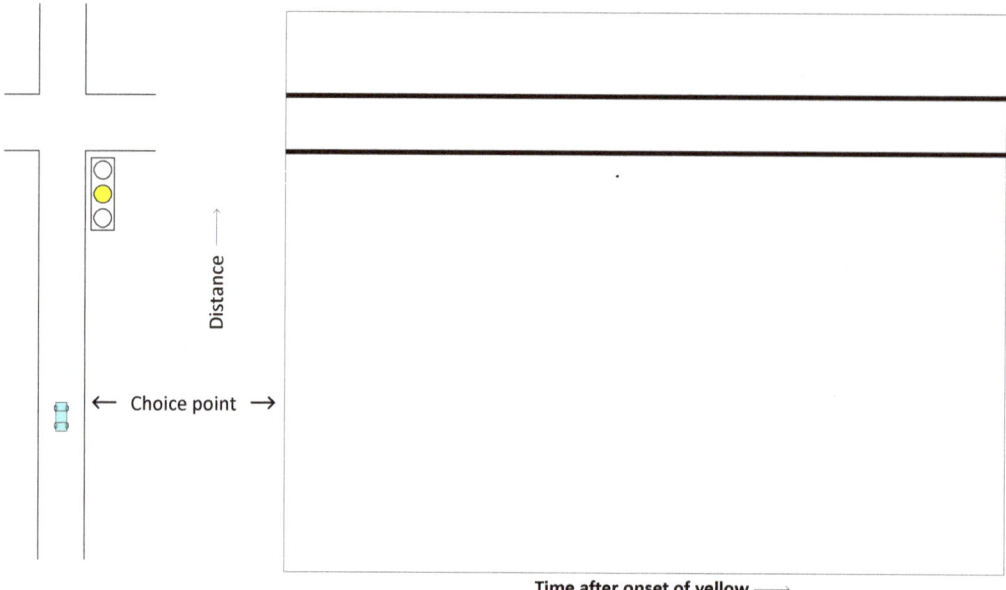

Figure 31. Vehicle at choice point at onset of yellow indication

The components of the *stopping distance* can be explained using Equation 1, which shows that this distance consists of two terms. The first term is the distance travelled during the perception-reaction time, which is the distance traveled while a driver perceives the indication change and then reacts to that change. The driver then initiates a braking maneuver; this braking distance is given by the second term of the equation.

Equation 1

$$x_s = v\delta + \frac{v^2}{2a}$$

where

x_s = distance from choice point to intersection stop bar, ft,
v = velocity of vehicle, ft/sec,
δ =perception-reaction time, sec, and
a = deceleration rate, ft/sec^2.

The stopping distance concept is illustrated in Figure 32. The stopping distance is composed of the perception-reaction distance $v\delta$ and the braking distance. The stopping time is composed of the perception-reaction time δ and the braking time. The vehicle trajectory (dashed line) shows the vehicle beginning to react to the yellow display at the choice point, then braking and coming to a complete stop at the stop bar.

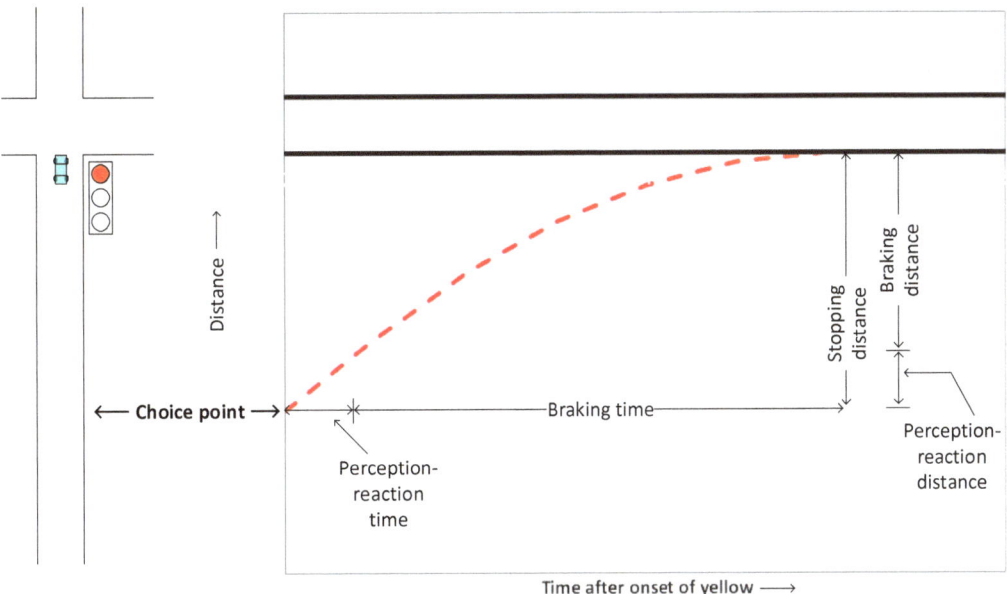

Figure 32. Distance from choice point to intersection stop bar, stopping distance

The other property of the choice point is that it is the farthest point upstream of the stop bar at which the driver can safely and completely clear the intersection, if the driver chooses to do so, when yellow is first displayed. Figure 33 shows a vehicle (solid line) following this trajectory, traveling from the choice point to a point at which the rear bumper of the vehicle is just downstream of the far side of the intersection. The components of this *clearing distance* are shown in the figure and are given by Equation 2.

Equation 2

$$x_c = x_s + w + L_v$$

where

    x_c = clearing distance, the distance from choice point to rear bumper clearing intersection, ft,

    x_s = stopping distance, or distance from choice point to intersection, ft,

    w = width of the intersection, ft, and

    L_v = length of the vehicle, ft.

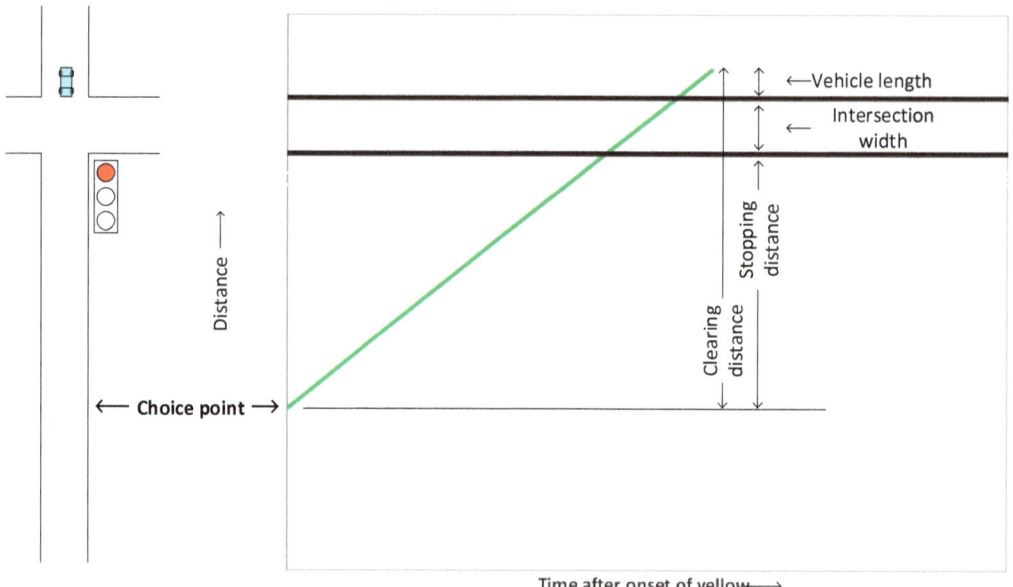

Figure 33. Clearing distance and components

4.2 Yellow and Red Clearance Times

The time that it takes for a vehicle to travel the clearing distance is set equal to the sum of the yellow and red clearance times, and is given in Equation 3.

Equation 3

$$Y + RC = \frac{x_c}{v} = \frac{x_s + w + L}{v} = \frac{x_s}{v} + \frac{w + L_v}{v}$$

where

Y = yellow time, sec,
RC = red clearance time, sec,
x_s = distance from choice point to intersection stop bar, ft,
x_c = clearing distance, ft,
v = velocity of vehicle approaching the intersection, ft/sec,
w = intersection width, ft, *and*
L_v = length of vehicle, ft.

The *yellow time* is set so that the vehicle can travel at a constant approach velocity from the choice point to the stop bar during the time that the yellow signal is displayed. Thus the yellow time is set equal to the first term of Equation 3, as shown in Equation 4.

Equation 4

$$Y = \frac{x_s}{v} = \delta + \frac{v}{2a}$$

where

Y = yellow time, sec,
x_s = distance from choice point to intersection stop bar, ft,
v = velocity of vehicle approaching the intersection, ft/sec,

δ = perception-reaction time, sec, and
a = deceleration rate, ft/sec^2.

The *red clearance time* is set so that the vehicle has time to travel from the stop bar to a point where the rear bumper just clears the intersection. Thus the red clearance time is set equal to the second part of Equation 3, as shown in Equation 5.

Equation 5

$$RC = \frac{w + L_v}{v}$$

where
    RC = red clearance time, sec,
    w = width of the intersection, ft,
    L_v = length of the vehicle, ft, and
    v = velocity of vehicle, ft/sec.

The relationship of the yellow and red clearance intervals to the trajectories of vehicles stopping at (dashed line) and clearing (solid line) the intersection is shown in Figure 34.

- A vehicle at the choice point that decides to stop has just enough distance to come to a complete stop at point A.
- A vehicle at the choice point that decides not to stop in response to the yellow display, sees the red displayed just as it passes the stop line. It reaches point B safely clearing the intersection during the red clearance time interval.

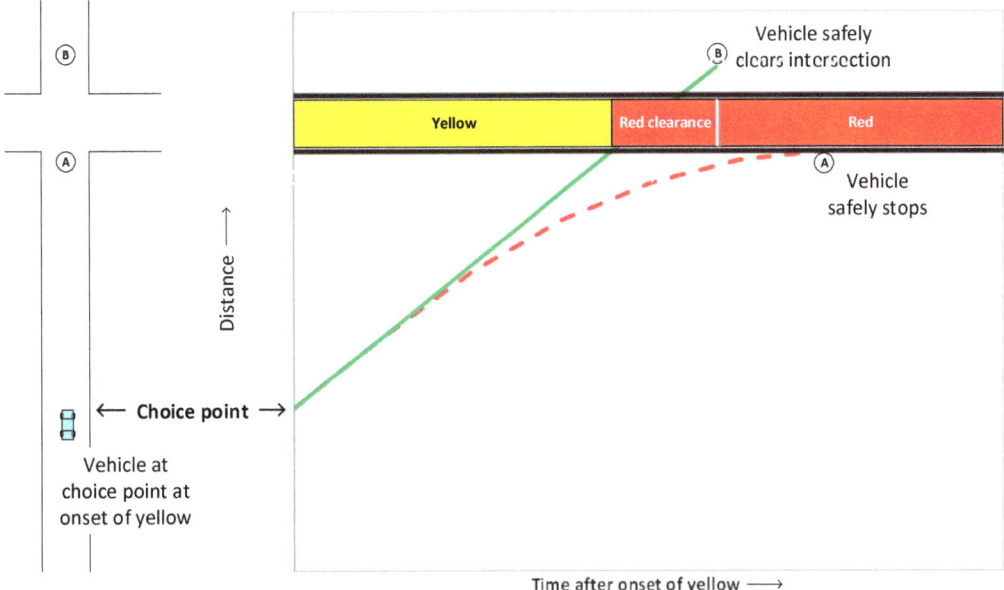

Figure 34. Yellow and red clearance intervals

A dilemma zone may result if any of the following conditions are true:

- The yellow and red clearance times used are shorter than the values computed using Equation 4 and Equation 5.
- The speed of the approaching vehicle is different than v.
- The length of a vehicle is greater than L_v.

A dilemma zone, in which the yellow time interval is less than required by Equation 4, is illustrated in Figure 35.

- Vehicle 1 is at the choice point when the yellow is first displayed. It is able to safely stop as shown by the dashed trajectory line.
- Vehicle 2 is at point A when yellow is first displayed. Shown by the solid trajectory line, it is at the farthest point upstream at which a vehicle can safely clear the intersection (point B) when the red clearance interval ends.
- Any vehicle located between vehicles 1 and 2 would be in the dilemma zone, and would neither be able to safely stop nor safely clear the intersection.

By using Equation 4 to compute the yellow time and Equation 5 to compute the red clearance time, a dilemma zone can be avoided. However, if the times used are longer than calculated from these equations, drivers will tend to take chances and enter the intersection when it is not safe to do so. Longer yellow times tend to encourage red-light running with the possibility of an increase in crashes with traffic from the side street.

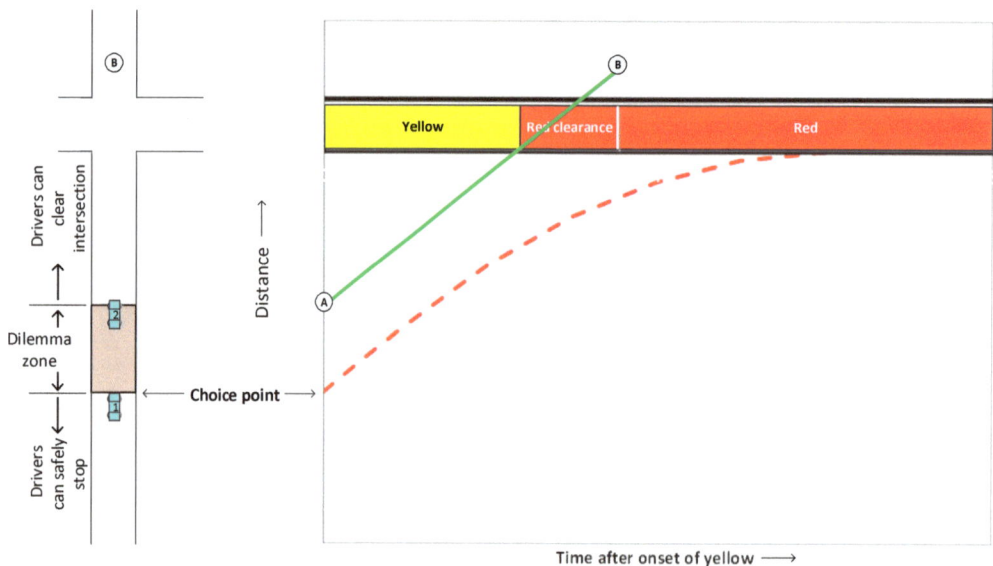

Figure 35. Illustration of dilemma zone with less than required yellow time

Example 7. Calculating the Yellow and Red Clearance Times
Consider an intersection with a speed limit of 35 mi/hr (51.3 ft/sec) and a width of 40 ft. Assume that a driver's perception reaction time is 1.0 sec. Also assume that the vehicle length is 20 ft and a comfortable deceleration rate is

10 ft/sec². What are the required yellow and red clearance interval times for the approach to this intersection?

Equation 4 is used to determine the duration of the yellow interval.

$$Y = \delta + \frac{v}{2a} = (1.0 \ sec) + \frac{51.3 \ ft/sec}{2(10 \ ft/sec^2)} = 3.6 \ sec$$

Equation 5 is used to determine the duration of the red clearance interval.

$$RC = \frac{w + L_v}{v} = \frac{40 \ ft + 20 \ ft}{51.3 \ ft/sec} = 1.2 \ sec$$

Notes:
- The minimum yellow time allowed by the MUTCD is 3.0 sec.
- The yellow and red clearance times are calculated to the nearest 0.1 second.

4.3 Summary of Section 4

What You Should Know and Be Able to Do:
- Describe the basis for determining the change and clearance intervals
- Calculate the duration of the yellow and red clearance times

Concepts You Should Understand:
- Concept 4.1: The choice point
  The choice point has two important properties. It is the closest point upstream of the intersection stop bar where the driver can safely stop at the onset of yellow. It is also the farthest point upstream of the stop bar at which the driver can safely and completely clear the intersection, if the driver chooses to do so, when yellow is first displayed.

- Concept 4.2: Duration of yellow and red clearance intervals
The yellow time is computed to be long enough so that the driver can travel from the choice point to the stop bar during the yellow display. The red clearance time is computed as the travel time from the stop bar to the point where the rear of the vehicle clears the intersection.

- Concept 4.3: The dilemma zone
In a dilemma zone, a driver can neither safely stop at the stop bar nor safely clear the intersection at the onset of yellow. A dilemma zone will result if the yellow and red clearance times used are less than those calculated as per the methods described in this section.

5. CAPACITY

Learning objectives:
- Define and apply the terms saturation flow rate, lost time, effective green time, and capacity
- Describe and apply the relationship between saturation flow rate, effective green time, and capacity

The notion of capacity is a common and important one in many fields of engineering. What load can be accommodated by a bridge? How many files can the hard drive on your computer hold? Most analyses of the capacity of a transportation system or highway facility focus on the peak hour. This is the time period when demand is at its highest point and when it may exceed the capacity of a system or facility.

The Highway Capacity Manual defines *capacity* as "...the maximum sustainable hourly flow rate at which persons or vehicles reasonably can be expected to traverse a point or a uniform section of a roadway during a given time period under prevailing roadway, environmental, traffic, and control conditions."[2] In short, for our purposes: capacity is the maximum rate of flow of vehicles or persons past a given point.

For a signalized intersection, the capacity of an approach is dependent on the control conditions present at the intersection. The control conditions that most directly affect capacity are the phasing and timing plans. This section describes the factors that need to be understood in order to determine the capacity of an intersection approach: the saturation flow rate, the lost time, and the effective green time.

5.1 Saturation Flow Rate

Suppose the signal display has just turned to green, and the vehicles that formed a queue during red begin to move into the intersection. There is some initial delay as the drivers respond to the green display. But soon vehicles are "up to speed" as they enter into and depart from the intersection. As long as the queue that formed during red continues to clear, the flow rate that we would observe at the stop line is called the *saturation flow rate*.

The saturation flow rate was shown earlier in this module (section 2) as part of the representation of traffic flow at a signalized intersection using a D/D/1 queuing model. In the flow profile diagram shown in Figure 7, the departure or service rate during the initial part of green was noted as the saturation flow rate. In Figure 10 (the cumulative vehicle diagram), the slope of the cumulative departure line during the initial portion of green is also equal to the saturation flow rate.

The Highway Capacity Manual suggests a base saturation flow rate of 1900 vehicles per hour per lane be used for traffic analysis. Typically, field

---

[2] Reference [3], p. 4-1.

measurements show a lower saturation flow rate due to constraining conditions of narrow lanes, on-street parking, the presence of heavy vehicles, and crossing pedestrians. Furthermore, if a left turn movement is permitted (and not protected), its saturation flow rate would be significantly lower since it must yield to opposing through and right turning vehicles.

The headway between vehicles departing at the saturation flow rate is called the *saturation headway*. This parameter was noted in the time-space diagram shown in Figure 4 (section 2 of this module). The relationship between the saturation headway and the saturation flow rate is shown in Equation 6.

Equation 6

$$h_s = \frac{3600 \ sec/hr}{s}$$

where

h_s = saturation headway, sec/veh, and

s = saturation flow rate, veh/hr.

5.2 Lost Time

Traffic streams are continuously starting and stopping at a signalized intersection as the right-of-way is transferred from one set of traffic movements to another. Each time one set of traffic movements is stopped, change and clearance intervals are provided to allow a time separation before the next set of movements are served. This means that a portion of the cycle cannot be completely utilized, translating to time that is lost to serving traffic. This time is called *lost time*, and includes both time at the beginning of green (start-up lost time) and time at the end of green (clearance lost time).

Startup lost time occurs when the signal indication turns from red to green and drivers in the queue do not instantly start moving at the saturation flow rate; there is an initial lag as drivers react to a change in the signal indication.

The concept of startup lost time is illustrated in Figure 36, which shows the headways between vehicles departing in the queue after the start of green. Typically, the headways for the first four vehicles exceed the saturation headway, as these vehicles react to the change in the signal indication and begin to move into the intersection. The difference between the actual headway and the saturation headway is the lost time for that vehicle. For example, h_1, the headway for the first vehicle in the queue, is the sum of its lost time t_1 and the saturation headway h_s.

The sum of the individual lost times for the first four vehicles is defined as the start-up lost time t_{sl}, as shown in Equation 7.

Equation 7

$$t_{sl} = \sum_{i=1}^{4} t_i$$

where

t_i = lost time for vehicle i, sec.

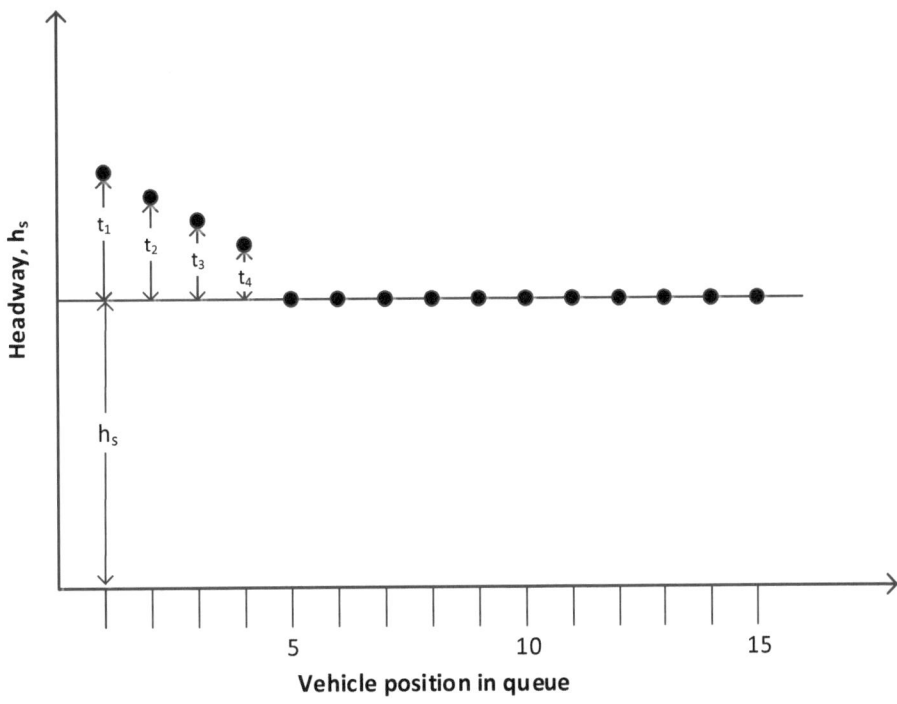

Figure 36. Saturation headway and start-up lost time

Lost time may also occur at the end of a phase. Some traffic may continue into the intersection, even during yellow. But the remainder of the yellow interval and all of the red clearance interval can't be effectively used by traffic and is referred to as *clearance lost time*.

Start-up and clearance lost times are summed to produce the total lost time for a phase as given by Equation 8.

Equation 8

$$t_L = t_{sl} + t_{cl}$$

where
    t_L = total lost time for a phase, sec,
    t_{sl} = start-up lost time, sec, and
    t_{cl} = clearance lost time, sec.

The start-up lost time typically has a value of 2 sec, while the clearance lost time also has a typical value of 2 sec. Thus a value of 4 sec is often used for the total lost time per phase.

As we will see in the next section, the lost time is important in determining how much green time is effectively available to serve traffic demand.

5.3 Effective Green Time and Effective Red Time

The green time available to serve traffic is called the *effective green time*. The effective green time for a given movement is the sum of the displayed green, yellow, and red clearance interval times minus the total lost time.

Equation 9

$$g = G + Y + RC - t_L$$

where
  g = effective green time, sec,
  G = displayed green time, sec,
  Y = yellow time, sec,
  RC = red clearance time, sec, and
  t_L = total lost time for a phase, sec.

The *effective red time* is the time during the cycle that is not available to serve traffic. The effective red time r is calculated as

Equation 10

$$r = C - g$$

where
  C = cycle length, sec, and
  g = effective green time, sec.

The relationship between the displayed green time G, the effective green time g, and the lost times (t_{sl} and t_{cl}) is represented graphically in Figure 37.

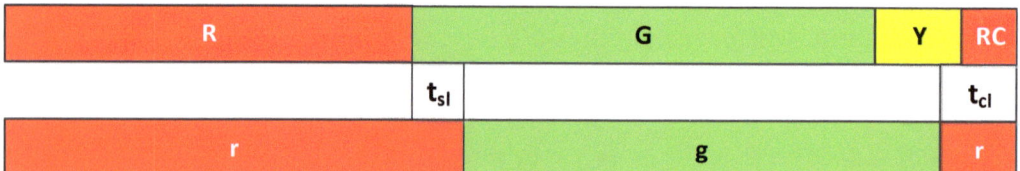

Figure 37. Relationship between displayed green time, effective green time, and lost time

5.4 Capacity of a Lane or an Approach

While the *saturation flow rate* is the maximum number of vehicles that can pass by the stop line on a lane or approach if the green indication is displayed continuously for an hour, the *capacity* is the maximum flow rate that would be observed based on the amount of green time that is actually available.

The proportion of the hour that is effectively available for a given movement is called the *effective green ratio*, or the ratio of the effective green time to the cycle length. The capacity of an approach or a lane is thus defined as the product of the saturation flow rate and the effective green ratio, as given in Equation 11.

5. Capacity

Equation 11

$$c = s \left(\frac{g}{C}\right)$$

where

c = capacity of an approach or lane, veh/hr,
s = saturation flow rate, veh/hr, and
g/C = effective green ratio (effective green time divided by cycle length).

Example 8. Determine Approach Capacity

For one approach of a signalized intersection (Figure 38), the saturation flow rate is 1900 vehicles per hour. The green is displayed for 15 sec, while the sum of the yellow and red clearance displays is 5 sec. The lost time is 4 sec. There are 60 cycles in one hour. What is the capacity of the approach?

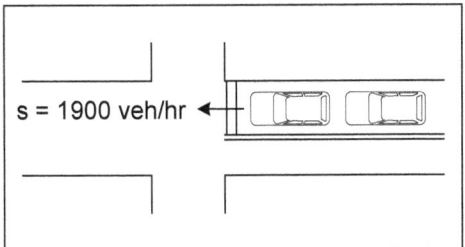

Figure 38. Intersection for Example 8

Since there are 60 cycles/hr, we know that the cycle length is:

$$C = \frac{(3600 \ sec/hr)}{(60 \ cycle/hr)} = 60 \ sec$$

The effective green time is computed using Equation 9:

$$g = G + Y + R_c - t_L = 15 \ sec + 5 \ sec - 4 \ sec = 16 \ sec$$

The effective green ratio is given by

$$\frac{g}{C} = \frac{16 \ sec}{60 \ sec} = 0.27$$

The capacity of the approach is calculated as the product of the saturation flow rate and the effective green ratio, using Equation 11.

$$c = 1900 \ veh/hr \ \times 0.27 = \ 513 \ veh/hr$$

5.5 Summary of Section 5

What You Should Know and Be Able to Do:

- Define and apply the terms saturation flow rate, lost time, effective green time, and capacity

- Describe and apply the relationship between saturation flow rate, effective green time, and capacity

Concepts You Should Understand:
- Concept 5.1: Saturation flow rate

The saturation flow rate is the maximum flow rate that we would observe for a queue departing from the intersection if the display was continuously green for an entire hour.

- Concept 5.2: Lost time and effective green time

The effective green time is the time available to serve traffic. Startup lost time is the time lost as the queue begins to move at the beginning of green. Clearance lost time is the portion of time at the end of the yellow (and all of the red clearance time) that is not available to serve traffic.

- Capacity 5.3: Capacity

The capacity is the maximum flow rate departing from the intersection during the hour considering the effective green time available to drivers. The capacity is less than the saturation flow rate because only part of the cycle is allocated to a movement.

6. SUFFICIENCY OF CAPACITY

> **Learning objectives:**
> - Compute and interpret measures of capacity sufficiency for intersection performance
> - Use the critical movement analysis to estimate intersection utilization

When designing a new intersection or evaluating the operation of an existing intersection, a common question to ask is: what is the capacity and is it sufficient to accommodate the traffic volume? A measure often used to determine whether there is sufficient capacity on an intersection approach is the volume-to-capacity ratio. A common method used to determine the volume-to-capacity ratio for an entire intersection is the critical movement analysis. Both the volume-to-capacity ratio and the critical movement analysis are described in this section.

6.1 Flow Ratio and Volume-to-Capacity Ratio

The *flow ratio* is defined as the ratio of the volume for a given movement to the saturation flow rate for that movement, or

Equation 12

$$Y = v/s$$

where
  Y = flow ratio,
  v = volume, veh/hr, and
  s = saturation flow rate, veh/hr.

The flow ratio is the proportion of an hour that is required to serve a traffic movement. Thus the flow ratio determines the minimum effective green ratio required to serve that movement.

Example 9. Flow Ratio and Effective Green Ratio
Suppose that the volume on one approach of a signalized intersection is 600 veh/hr and the saturation flow rate for the approach is 1900 veh/hr. What proportion of the hour should be made available to serve this movement so that sufficient capacity is provided?

Using Equation 12, the flow ratio Y is determined to be

$$Y = \frac{600 \; veh/hr}{1900 \; veh/hr} = 0.32$$

This means that the effective green ratio must be at least 0.32 if sufficient capacity is to be provided to serve the demand on this approach.

The *volume-to-capacity ratio* is defined in Equation 13.

Equation 13

$$X = v/c$$

where
   X = volume-to-capacity ratio,
   v = volume, veh/hr, and
   c = capacity, veh/hr.

Since the capacity is the product of the saturation flow rate and the effective green ratio, we can rewrite the definition of the volume-to-capacity ratio as

Equation 14

$$X = \frac{v}{s\left(\frac{g}{C}\right)} = \frac{v/s}{g/C} = \frac{Y}{g/C}$$

where
   v = volume, veh/hr,
   s = saturation flow rate, veh/hr,
   g = effective green time, sec, and
   C = cycle length, sec.

We can see from Equation 14 that if the flow ratio is less than the effective green ratio, then the volume-to-capacity ratio will be less than one. In this case, there will be sufficient capacity to serve the traffic demand. However, if the flow ratio is greater than the effective green ratio, there will not be sufficient capacity to serve the demand.

Example 10. Volume-to-Capacity Ratio
Suppose that the volume on an intersection approach is 750 veh/hr while the saturation flow rate is 1900 veh/hr. If the effective green ratio is 0.42, what is the volume-to-capacity ratio for the approach?

We first compute the capacity of the approach:

$$c = s\left(\frac{g}{C}\right) = 1900 \; veh/hr \; (0.42) = 798 \; veh/hr$$

The volume-to-capacity ratio is then:

$$X = \frac{v}{c} = \frac{750 \; veh/hr}{798 \; veh/hr} = .94$$

6. Sufficiency of Capacity

There are several interpretations of this result that are worth noting. First, there is just enough capacity to serve the demand, as *X* is just under one. However, if there is any variation in the demand (even small variations from one cycle to the next), the demand may exceed capacity. So, as we will see in the next section, it is prudent to provide some margin of safety in the design. Often a design value used for the volume-to-capacity ratio is 0.85 or less.

6.2 Critical Movement Analysis

While the definitions for volume-to-capacity ratio and flow ratio apply to an individual movement, we often want to determine the volume-to-capacity ratio for the entire intersection. If we know the volume-to-capacity ratio for the entire intersection, we can answer the question that began this section: is there sufficient capacity at the intersection to accommodate the traffic volume existing at or projected for the intersection?

The critical movement analysis method is commonly used for answering this question. The critical movement analysis method includes the five steps described below. The method is then illustrated with two examples, one for protected left turns and one for permitted left turns.

Step 1: Compute the flow ratio Y_i for each movement i present at the intersection. The standard movement numbers and notations are shown in Figure 39. Right turn volumes are combined with the through movement volumes.

Equation 15

$$Y_i = \frac{v_i}{s_i}$$

where
Y_i = flow ratio for movement i
v_i = volume for movement i, veh/hr, and
s_i = saturation flow rate for movement i, veh/hr.

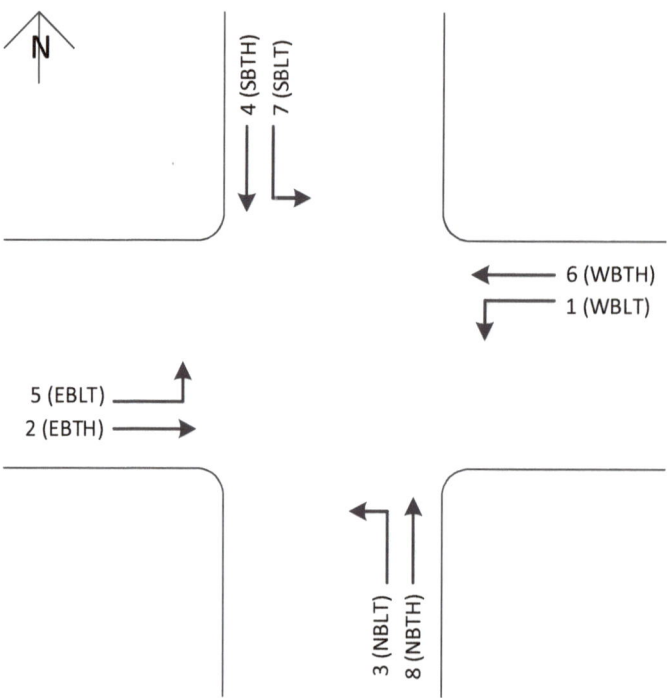

Figure 39. Numbering and notation of movements at a signalized intersection

Step 2: Determine the flow ratio sums for the phase sequences in each ring for each concurrency group (for the case of protected left turns only). (For permitted left turns, skip to step 3.)

Figure 40 and Figure 41 show the movements and phases from the east-west concurrency group.

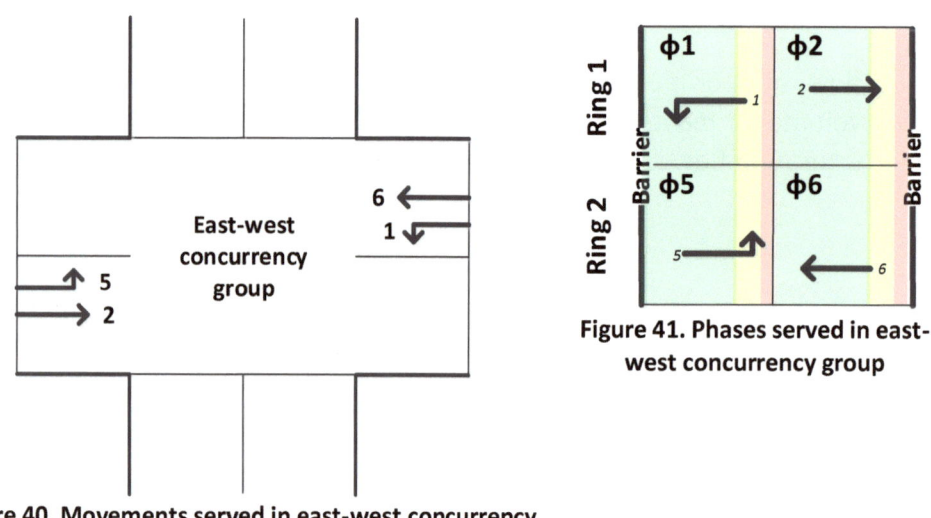

Figure 40. Movements served in east-west concurrency group

Figure 41. Phases served in east-west concurrency group

For the east-west concurrency group, the flow ratio sums for the movements served in rings 1 and 2 are given by Equation 16 and Equation 17.

6. Sufficiency of Capacity

Equation 16

$$Y_{EW1} = Y_1 + Y_2$$

Equation 17

$$Y_{EW2} = Y_5 + Y_6$$

where
   Y_{EW1} = flow ratio sum for ring 1,
   Y_{EW2} = flow ratio sum for ring 2, and
   Y_i = flow ratios for movements 1, 2, 5, and 6

Figure 42 and Figure 43 show the movements and phases from the north-south concurrency group.

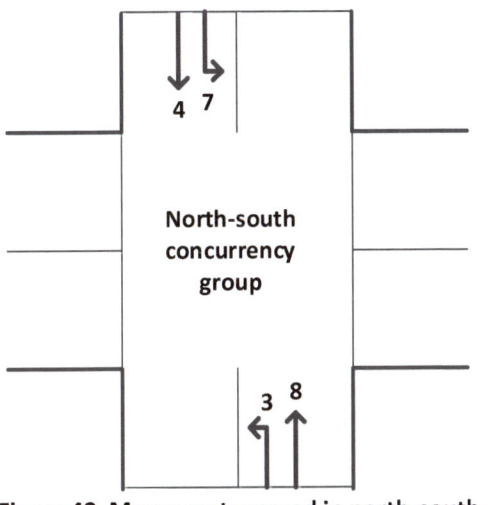

Figure 42. Movements served in north-south concurrency group

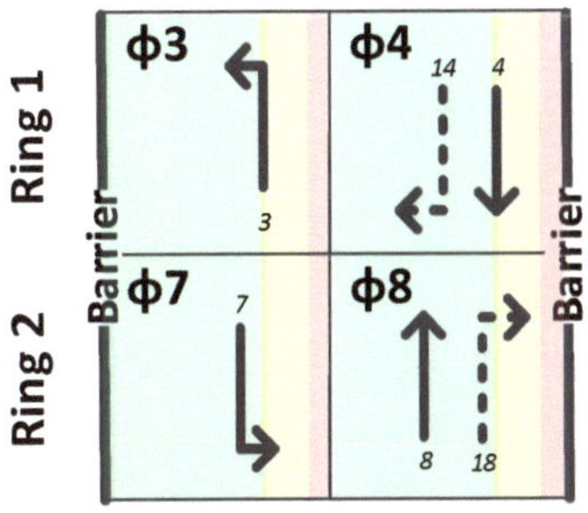

Figure 43. Phases served in north-south concurrency group

For the north-south concurrency group, the flow ratio sums are given by Equation 18 and Equation 19.

Equation 18

$$Y_{NS1} = Y_3 + Y_4$$

Equation 19

$$Y_{NS2} = Y_7 + Y_8$$

where
   Y_{NS1} = flow ratio sum for ring 1,
   Y_{NS2} = flow ratio sum for ring 2, and
   Y_i = flow ratios for movements 3, 4, 7, and 8.

Step 3: Within each concurrency group, identify the movements with the maximum flow ratio sum (for protected left turns) or the movement with the

maximum flow ratio (for permitted left turns). These movements are the critical movements for each concurrency group.

For protected left turns, we use Equation 20 and Equation 21 to determine the critical movements for the east-west and north-south concurrency groups.

Equation 20

$$Y_{EW-critical} = Max(Y_{EW1}, Y_{EW2})$$

where

$Y_{EW-critical}$ = critical flow ratio for the EW concurrency group,
Y_{EW1} = flow ratio sum for ring 1, and
Y_{EW2} = flow ratio sum for ring 2.

Equation 21

$$Y_{NS-critical} = Max(Y_{NS1}, Y_{NS2})$$

where

$Y_{NS-critical}$ = critical flow ratio for NS concurrency group,
Y_{NS1} = flow ratio sum for ring 1, and
Y_{NS2} = flow ratio sum for ring 2.

For permitted left turns, we use Equation 22 and Equation 23 to determine the critical movements for each concurrency group.

Equation 22

$$Y_{EW-critical} = Max(Y_1, Y_2, Y_5, Y_6)$$

where

$Y_{EW-critical}$ = critical flow ratio for the EW concurrency group, and
Y_i = flow ratios for movements 1, 2, 5, and 6.

Equation 23

$$Y_{NS-critical} = Max(Y_3, Y_4, Y_7, Y_8)$$

where

$Y_{NS-critical}$ = critical flow ratio for NS concurrency group, and
Y_i = flow ratios for movements 3, 4, 7, and 8.

Step 4: Determine the critical volume-to-capacity ratio for the intersection.

Step 4a: Compute the lost time per cycle L as given in Equation 24.

6. Sufficiency of Capacity

Equation 24

$$L = \sum_{i=1}^{M} t_{Li}$$

where

L = lost time per cycle, sec,

t_{Li} = lost time for phase i, and

M = number of phases that serve the critical movements for one cycle.

If all left turn movements are protected, M is 4; if all are permitted, M is 2. If the left turn movements for one concurrency group are protected and if they are permitted for the other concurrency group, M is 3.

Step 4b: Compute the *critical volume-to-capacity* ratio considering the critical flow ratios and the lost time per cycle, using Equation 25.

Equation 25

$$X_c = \frac{(Y_{EW-critical} + Y_{NS-critical})(C)}{C - L}$$

where

X_c = critical volume-to-capacity ratio,

$Y_{EW-critical}$ = critical flow ratio for the EW concurrency group,

$Y_{NS-critical}$ = critical flow ratio for the NS concurrency group,

C = cycle length, sec, and

L = total lost time per cycle for the critical phases, sec.

Step 5: Based on the value of X_c calculated in step 4, determine the sufficiency of capacity.

Table 4 gives four possible sufficiency ratings based on the critical volume-to-capacity ratio computed for the intersection. It is desirable, though not often possible during the peak period, for X_c to be less than 0.85.

Table 4. Sufficiency of capacity

X_c	Sufficiency of capacity rating
< 0.85	Intersection is operating under capacity. Excessive delays are not experienced.
0.85–0.95	Intersection is operating near its capacity. Higher delays may be expected, but continuously increasing queues should not occur.
0.95–1.00	Unstable flow results in a wide range of delay. Intersection improvements will be required soon to avoid excessive delays.
> 1.00	The demand exceeds the available capacity of the intersection. Excessive delays and queuing are anticipated.

Example 11. Critical Movement Analysis for Protected Left Turns

A standard 4-approach intersection has the geometric and volume characteristics shown in Figure 44. Figure 45 shows the phasing plan for this intersection, including leading protected left turns. The cycle length is 90 sec and the lost time is 4 sec/phase. The saturation flow rate is 1900 veh/hour/lane for through movements and protected left turn movements. Use the critical movement analysis method to determine whether the capacity is sufficient to serve the volume for this intersection.

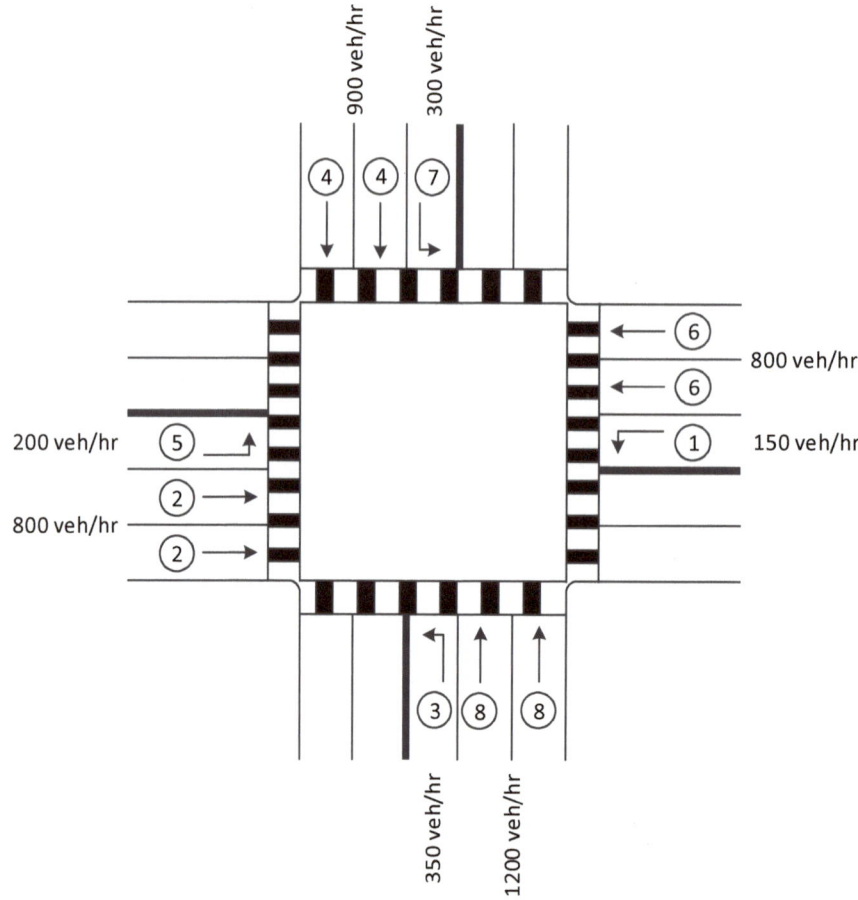

Figure 44. Intersection characteristics for Example 11

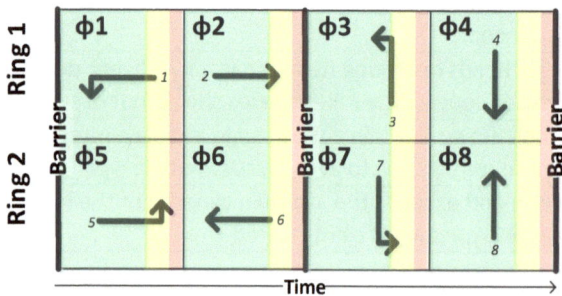

Figure 45. Phasing plan for Example 11

Step 1: Compute the flow ratio Y for each movement present at the intersection.

The flow ratios for each of the movements are calculated and shown in Figure 46.

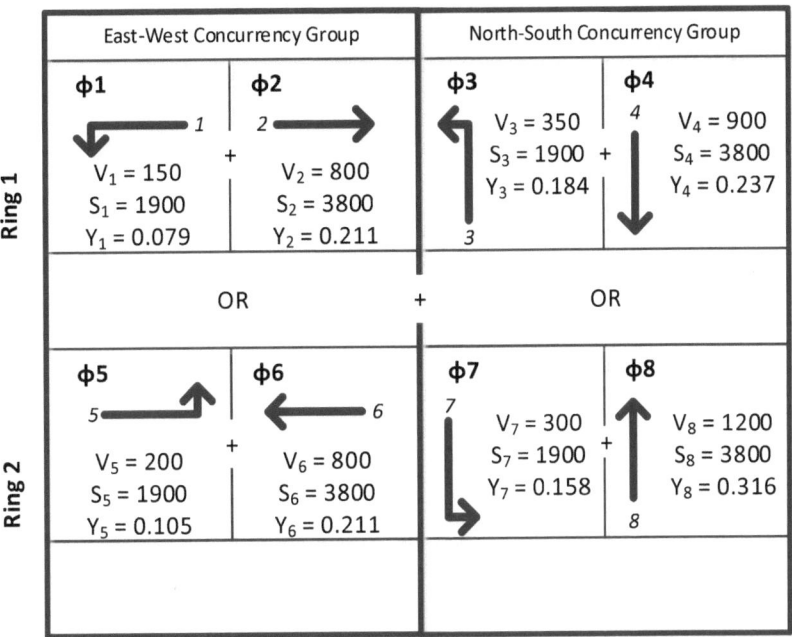

Figure 46. Flow ratios for each movement for Example 11

Step 2: Determine the flow ratio sums for the phase sequences in each ring for each concurrency group (for the case of protected left turns only).

The flow ratio sums for each ring within each concurrency group are shown in Figure 47. For example, for ring 1 in the east-west concurrency group, the flow ratios for movements 1 and 2 are .079 and .211, respectively. Their sum, noted as Y_{EW1}, is .290.

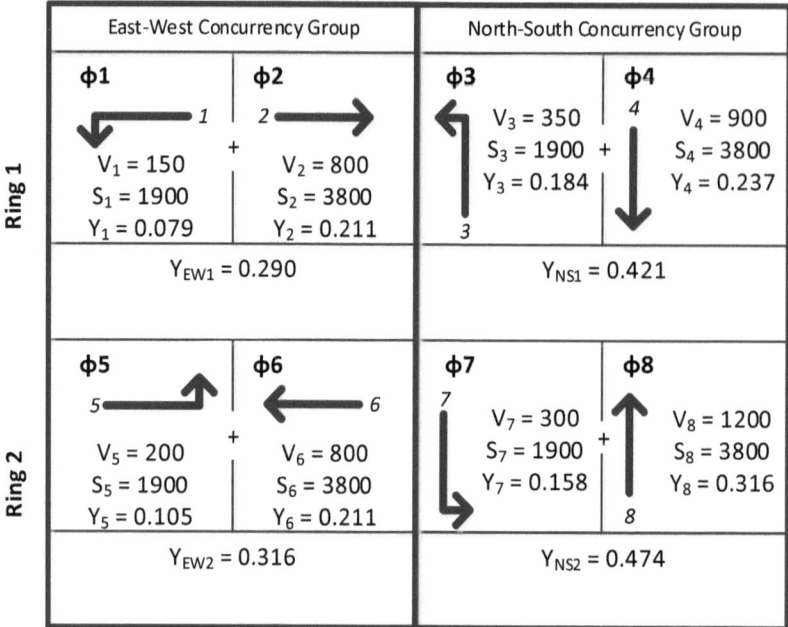

Figure 47. Flow ratio sums for Example 11

Step 3: Within each concurrency group, identify the movements with the maximum flow ratio sum (for protected left turns) or the movement with the maximum flow ratio (for permitted left turns). These movements are the critical movements for each concurrency group.

Since these are protected left turns, we identify the movements with the maximum flow ratio sum within each concurrency group (See Figure 48). For the east-west concurrency group, the movements served in ring 2 (movements 5 and 6) have the highest flow ratio sum (.316), as compared to the movements served in ring 1 (.290). For the north-south concurrency group, the movements served in ring 2 have the highest flow ratio sum (.474).

$$Y_{EW-critical} = Max(Y_{EW1}, Y_{EW2}) = Max(.290, .316) = .316$$
$$Y_{NS-critical} = Max(Y_{NS1}, Y_{SW2}) = Max(.421, .474) = .474$$

6. Sufficiency of Capacity

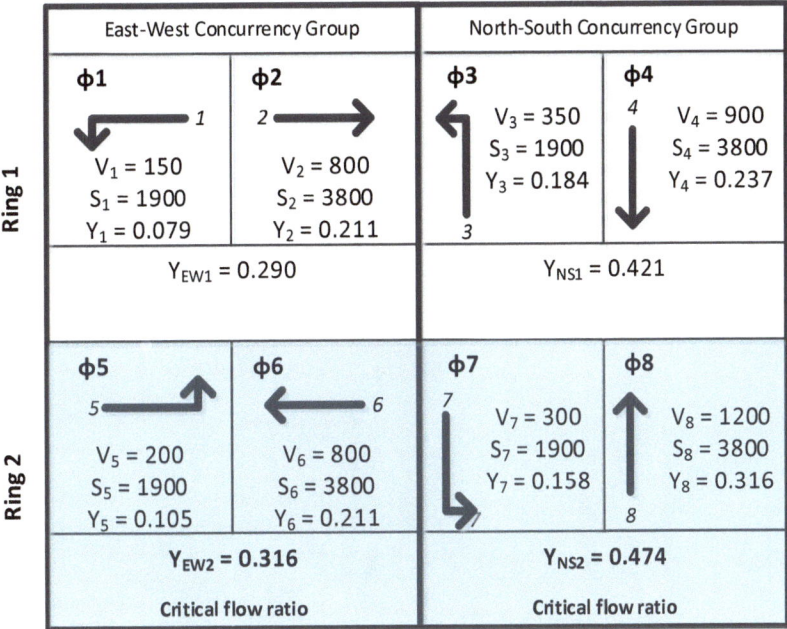

Figure 48. Critical flow ratios for Example 11

Step 4: Determine the critical volume-to-capacity ratio X_c for the intersection.

The critical flow ratios (Y_{EW2}=.316, Y_{NS2}=.474) were computed in step 3. The cycle length is given as 90 sec and the lost time per phase is 4 seconds. There are four critical phases since all left turns are protected, so the total lost time L is 16 sec. The critical volume-to-capacity ratio is computed using Equation 25:

$$X_c = \frac{(Y_{EW-critical} + Y_{NS-critical})(C)}{C - L} = \frac{(.316 + .474)(90\ sec)}{90\ sec - 16\ sec} = .96$$

Step 5: Based on the value of X_c calculated in step 4, determine the sufficiency of capacity.

The critical volume-to-capacity ratio, X_c = .96, indicates that the intersection is operating in the region of unstable flow and that excessive delays and queuing will result, using the ratings from Table 4.

Example 12. Critical Movement Analysis for Permitted Left Turns

A standard four-approach intersection has the geometric and volume characteristics shown in Figure 49. The phasing scheme is shown in Figure 50. The cycle length is 90 sec and the lost time is 4 sec per phase. The permitted left turns have a saturation flow rate of 450 veh/hr while through movements have a saturation flow rate of 1900 veh/hr.

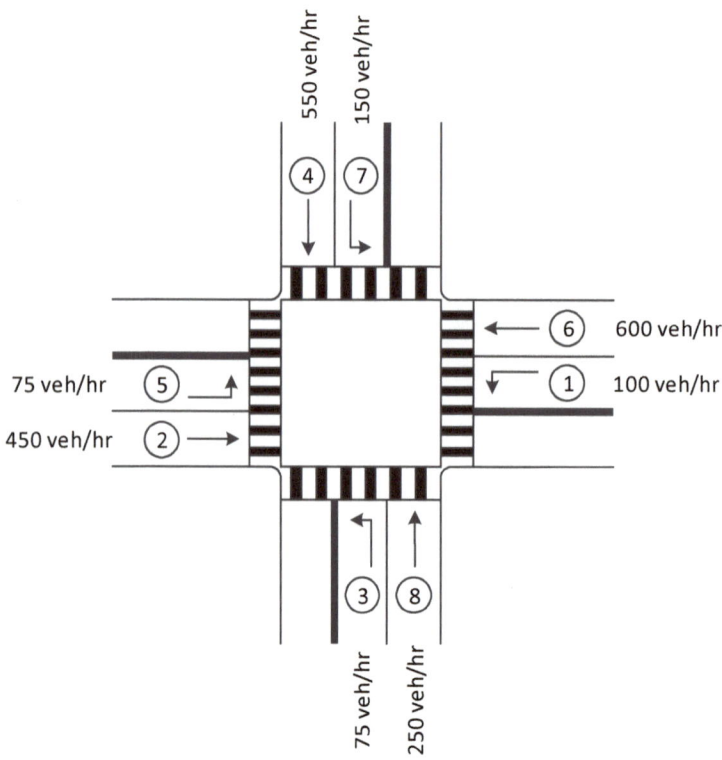

Figure 49. Intersection characteristics for Example 12

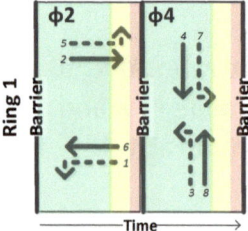

Figure 50. Ring barrier diagram for Example 12

Step 1: Compute the flow ratio Y$_i$ for each movement i present at the intersection.

The flow ratios for each of the movements are calculated and shown in Figure 51.

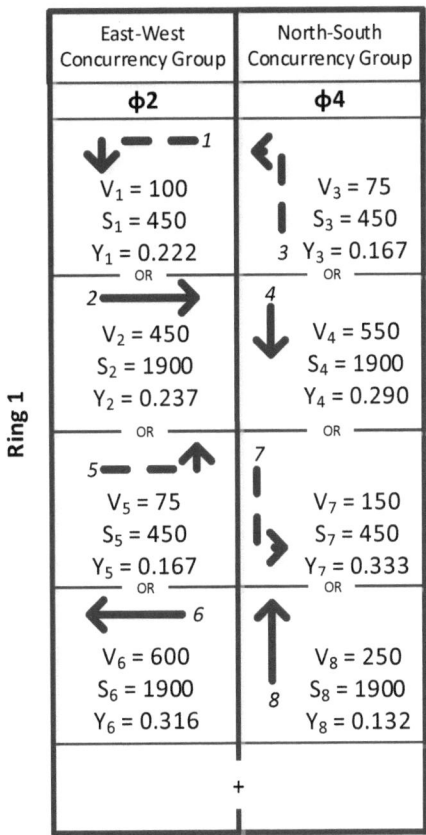

Figure 51. Flow ratios for each movement for, Example 12

Step 2: Determine the flow ratio sums for the phase sequences in each ring for each concurrency group (for the case of protected left turns only).
Since this example is for permitted left turns, we skip to step 3.

Step 3: Within each concurrency group, identify the movements with the maximum flow ratio sum (for protected left turns) or the movement with the maximum flow ratio (for permitted left turns). These movements are the critical movements for each concurrency group.
Since this example is for permitted left turns, we identify the movement with the maximum flow ratio in each concurrency group. As shown in Figure 52, movement 2 has the highest flow ratio (.316) for the east-west concurrency group. For the north-south concurrency group, movement 7 has highest flow ratio (.333).

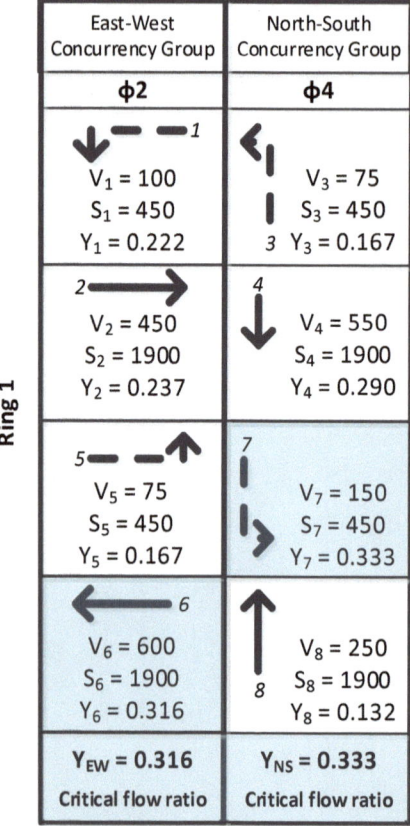

Figure 52. Critical flow ratios for Example 12

Step 4: Determine the critical volume-to-capacity ratio (X_c) for the intersection. The critical flow ratios (Y_{EW}=.316, Y_{NS}=.333) were determined in step 3. The cycle length is given as 90 sec and the lost time per phase is 4 sec. There are two critical phases since all left turns are permitted, so the total lost time L is 8 sec. The critical volume-to-capacity ratio is computed using Equation 25.

$$X_c = \frac{(Y_{EW-critical} + Y_{NS-critical})(C)}{C - L} = \frac{(.316 + .333)(90 \; sec)}{90 \; sec - 8 \; sec} = .71$$

Step 5: Based on the value of X_c calculated in step 4, determine the sufficiency of capacity.
The critical volume-to-capacity ratio, X_c = 0.71, indicates that the intersection is operating under capacity, using the ratings from Table 4.

6. Sufficiency of Capacity

6.3 Summary of Section 6

What You Should Know and Be Able to Do:

- Compute and interpret measures of capacity sufficiency for intersection performance
- Use the critical movement analysis to estimate intersection utilization

Concepts You Should Understand:

- Concept 6.1: Flow ratio and green ratio

The flow ratio shows the proportion of the hour needed to serve a traffic movement. The green ratio shows the proportion of an hour available to serve a traffic movement.

- Concept 6.2: Relationship between green ratio, flow ratio, and sufficiency of capacity

As long as the green ratio is greater than the flow ratio, there is sufficient capacity to serve the traffic demand.

- Concept 6.3: Critical movements in a concurrency group

For a concurrency group, the critical movements are those movements with the highest flow ratio sum considering each of the rings in the concurrency group (for protected left turns) or the movement with the highest flow ratio (for permitted left turns).

- Concept 6.4: Critical volume-to-capacity ratio

The critical volume-to-capacity ratio is the proportion of the available capacity at the intersection used by the critical traffic movements.

7. DELAY AND LEVEL OF SERVICE

> **Learning objectives**:
> - Calculate uniform delay at a signalized intersection
> - Describe and apply measures of effectiveness for a signalized intersection
> - Describe and apply the level of service framework

Just as we earlier asked the question "is there sufficient capacity at this intersection to accommodate demand", we can also ask another related question: "how is the intersection performing?" Or, stated another way: "what quality of service is provided to the users of the intersection?" Sometimes this question is asked for one intersection and its individual movements. Often it is asked when we are comparing alternative designs for an intersection or comparing the operation of several different intersections. Three performance measures are commonly used to evaluate intersection performance: delay, volume-to-capacity ratio, and back of queue size. Delay is a measure that can be perceived by users; it is used to determine level of service.

7.1 Determining Uniform Delay Using the Cumulative Vehicle Diagram and the Queue Accumulation Polygon

In section 2, the cumulative vehicle diagram and the queue accumulation polygon were presented as two ways of representing traffic flow at a signalized intersection. Figure 53 shows a cumulative vehicle diagram highlighting the cumulative number of vehicles that have arrived at and departed from the signalized intersection over time. The slope of the cumulative arrival line (solid line) is equal to the arrival rate v. Three time periods are shown for the cumulative departure line:
- Effective red (period 1), during which the departure flow is zero,
- The queue service time g_s (period 2), in which the slope of the cumulative departure line is equal to the saturation flow rate s, and
- The final portion of effective green (period 3), in which the slopes of the cumulative arrival line and the cumulative departure line are equal to the arrival rate v.

The horizontal line connecting the arrival line and the departure line for each vehicle (shown in Figure 53 as d_i) is the delay experienced by that vehicle. The length of the queue, measured in vehicles, is the vertical distance at a given point in time between the arrival line and the departure line. The area of the triangle is equal to the total delay experienced by all vehicles that arrive during the cycle. The delay is called uniform delay since vehicles are assumed to arrive at the intersection at a uniform flow rate.

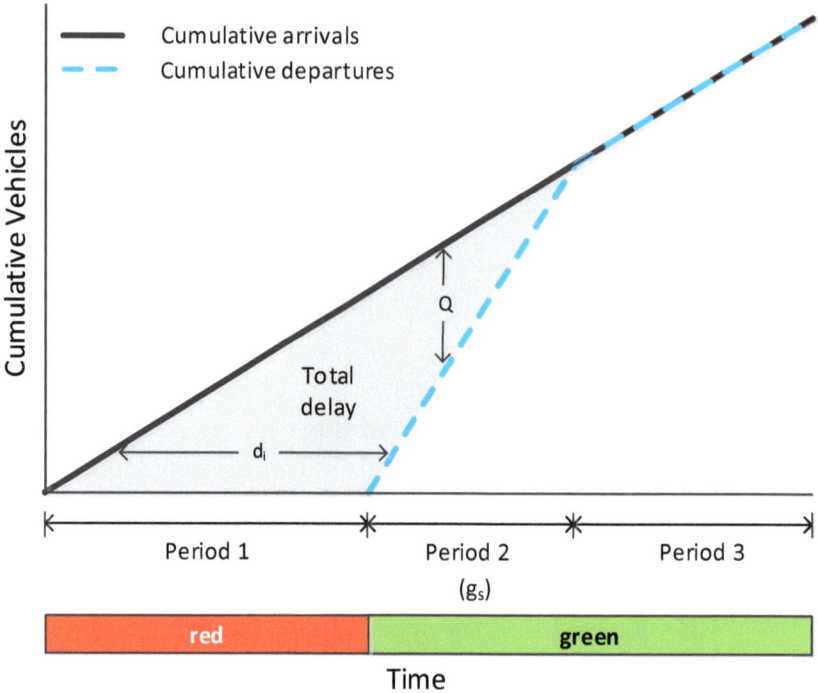

Figure 53. Cumulative vehicle diagram

Figure 54 shows a queue accumulation polygon for the same traffic flow conditions and time periods as shown in Figure 53. Here, the height of the polygon is the length of the queue at any point in time. The maximum queue length occurs at the end of the effective red interval. The area of the triangle is equal to the total delay experienced by all vehicles that arrive during the cycle.

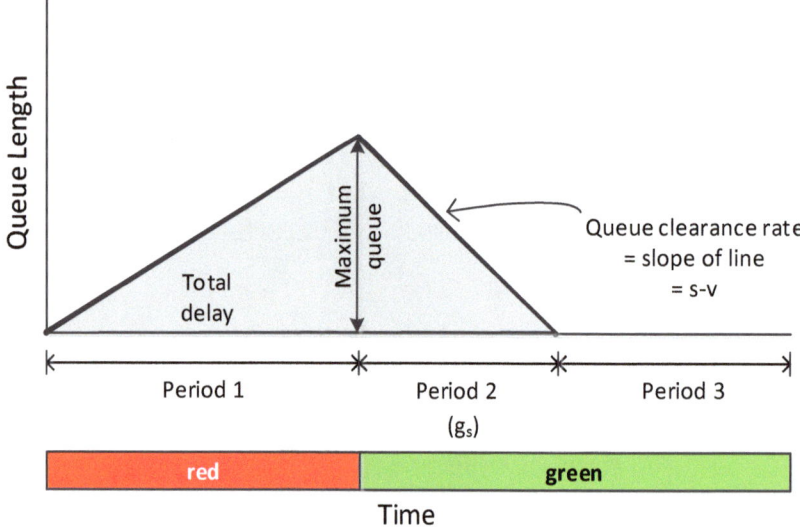

Figure 54. Queue accumulation polygon

Two assumptions made about the D/D/1 queuing model are represented in these diagrams:

7. Delay and Level of Service

- The queue clears before the end of the effective green, implying that the arrival volume is less than the capacity. This also implies that the queue at the beginning of red is zero.
- The arrival pattern is uniform.

The first step in determining the area of the triangle (total delay) is to compute the time that it takes for the queue to clear after the beginning of effective green. We call this time the queue service time g_s. When the queue clears, the number of vehicles that have arrived at the intersection since the beginning of the cycle must equal the number of vehicles that have departed since the beginning of the effective green. We can write this equality as

Equation 26

$$v(r + g_s) = sg_s$$

where
v = arrival rate, veh/sec,
r = effective red time, sec,
g_s = queue service time, sec, and
s = saturation flow rate, veh/sec.

Solving for g_s:

Equation 27

$$g_s = \frac{vr}{s - v}$$

Stated in words, the queue service time is equal to the length of the queue at the end of effective red (vr) divided by the rate of queue clearance after the start of effective green (s-v).

The area of the triangle in Figure 53 is equal to one-half the product of the effective red time (the base of the triangle) and the number of vehicles that have arrived at the intersection at the point that the queue has cleared, $v(r + g_s)$. This latter number is the height of the triangle. The total uniform delay D_t is given in Equation 28.

Equation 28

$$D_t = (0.5)(r)[(v)(r + g_s)]$$

where the variables are defined as above. Substituting g_s from Equation 27, we get another expression for the total delay.

Equation 29

$$D_t = (0.5)(r)\left[(v)\left(r + \frac{vr}{s - v}\right)\right]$$

The number of vehicles that arrive during the cycle is the product of the arrival rate v and the cycle length C. The *average uniform delay per vehicle* d_{avg} is the total delay from Equation 29 divided by the number of vehicles that arrive during the cycle (vC).

Equation 30

$$d_{avg} = \frac{(0.5)(r)}{vC}\left[(v)\left(r + \frac{vr}{s-v}\right)\right]$$

When terms are rearranged and simplified:

Equation 31

$$d_{avg} = (0.5r)\left[\frac{(1 - g/C)}{(1 - v/s)}\right]$$

The same equation will result if we compute the area from the queue accumulation polygon in Figure 54. These equations yield the average delay experienced by vehicles if the arrival pattern is uniform and if the demand is less than the capacity of the approach. When we examine the equation for average delay, we can see that delay increases when:

- The effective red increases, resulting in a longer queue that forms during red.
- The effective green ratio decreases, providing less green time during the cycle to serve the queue.
- The flow ratio increases, as the volume approaches the capacity.

The average uniform delay for the entire intersection can also be computed. If the delay for each approach is computed to be d_i and the volume on each approach is v_i, the average delay for the intersection d_{int} is the weighted average of the delays for each of the intersection approaches.

Equation 32

$$d_{int} = \frac{\sum d_i v_i}{\sum v_i}$$

where
d_i = delay for each of the i approaches at the intersection, sec, and
v_i = the volume for each of these approaches, veh/sec.

Example 13. Calculation Of Average Delay When Volume Is Less Than Capacity

An intersection approach has an arrival rate of 630 veh/hr and a saturation flow rate of 1900 veh/hr. The cycle length is 100 sec, the effective red time is 60 sec, and the effective green time is 40 sec. Determine the queue service time and the average delay for this approach.

Step 1. Convert the flow rates from veh/hr to veh/sec.

$$v = \frac{630\ veh/hr}{3600\ sec/hr} = .175\ veh/sec$$

$$s = \frac{1900\ veh/hr}{3600\ sec/hr} = .528\ veh/sec$$

Step 2. Using Equation 11, calculate the approach capacity and compare it to the arrival flow rate.

$$c = s \times \left(\frac{g}{C}\right) = 1900\ veh/hr\ \times \left(\frac{40\ sec}{100\ sec}\right) = 760\ veh/hr$$

The arrival volume (630 veh/hr) is thus less than the capacity (760 veh/hr), so the analytical method (Equation 31) can be used to calculate the average uniform delay.

Step 3. Calculate the queue service time using Equation 27.

$$g_s = \frac{vr}{s-v} = \frac{(.175\ veh/sec)(60\ sec)}{.528\ veh/sec - .175\ veh/sec} = 29.7\ sec$$

Step 4. Construct the cumulative vehicle diagram and the queue accumulation polygon. (Though this step is not necessary to solve this problem, the preparation of the two diagrams is recommended for better understanding these concepts).

Cumulative vehicle diagram: The arrival line (solid line, Figure 55) shows the cumulative number of vehicles that arrive from the beginning of the cycle (t=0) to the end of the cycle (t=100). The departure line (dashed line) shows the cumulative number of departures over time. During the effective red (from t=0 to t=60), the number of departures is zero. At the end of the effective red, the queue is 10.5 vehicles (line 1-2). From the beginning of effective green to the time that the queue clears (point 3), the departure line has a slope equal to the saturation flow rate. After the queue has cleared (t=89.7, 29.7 sec after the beginning of effective green), the arrival line and the departure line are coincident.

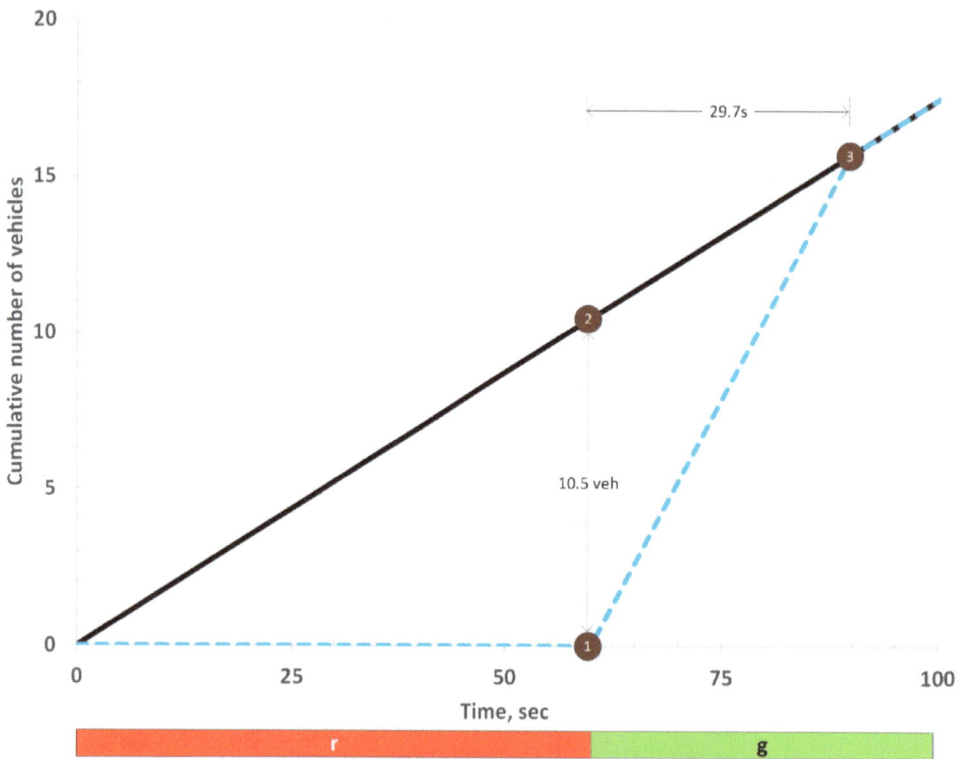

Figure 55. Cumulative vehicle diagram for Example 13

Queue accumulation polygon: The height of the polygon in Figure 56 shows the number of vehicles in the queue at any point during the cycle.

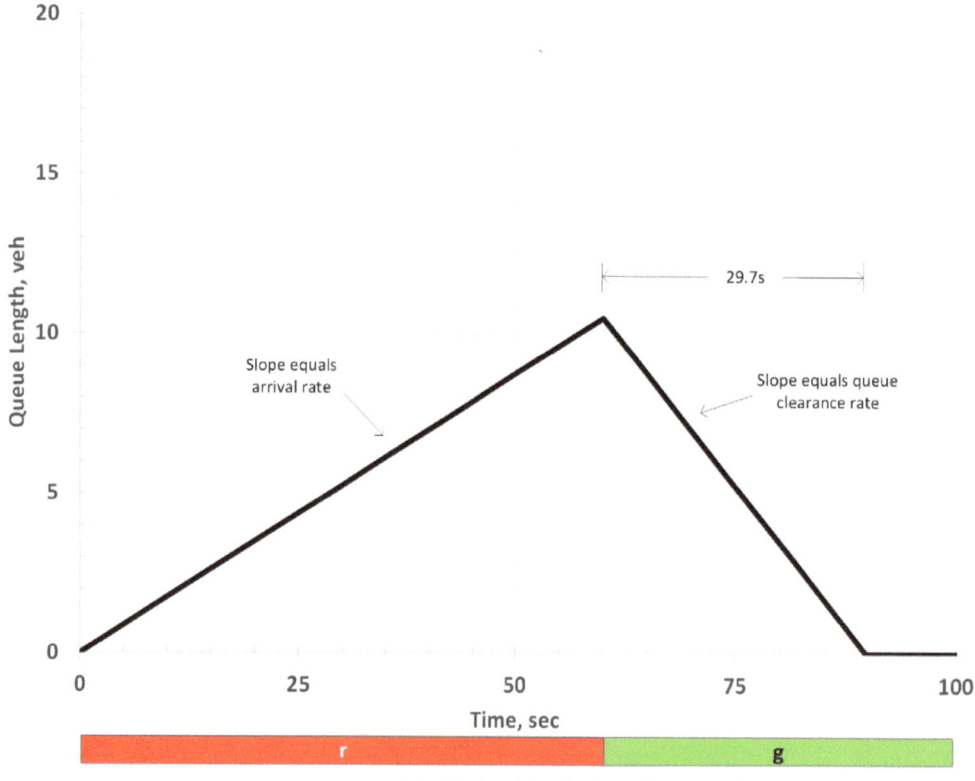

Figure 56. Queue accumulation polygon for Example 13

7. Delay and Level of Service

Step 5. The average delay is calculated using Equation 31:

$$d_{avg} = (0.5r) \frac{\left[\left(1 - \frac{g}{C}\right)\right]}{\left[\left(1 - \frac{v}{s}\right)\right]}$$

$$d_{avg} = (0.5 \times 60 \; sec) \left[\frac{\left(1 - \frac{40 \; sec}{100 \; sec}\right)}{\left[\left(1 - \frac{630 \; veh/hr}{1900 \; veh/hr}\right)\right]}\right] = 26.9 \; sec$$

When the volume exceeds the capacity, a basic assumption of the D/D/1 queuing model is violated and we must use the graphical method to compute the delay. The graphical method is based on finding the area of the triangle in either the cumulative vehicle diagram or the queue accumulation polygon. An example of the graphical method is given in Example 14.

Example 14. Calculation Of Average Delay When Volume Exceeds Capacity
An intersection approach has a cycle length of 100 sec and an effective green time of 40 sec. The arrival rate varies over three cycles. The arrival rate is 900 veh/hr during the first cycle, 720 veh/hr during the second cycle, and 540 veh/hr during the third cycle. Calculate the average delay for the approach over the three cycles. The saturation flow rate is 1900 veh/hr.

Step 1. Convert the flow rates from veh/hr to veh/sec. The resulting rates for each cycle are shown in Table 5.

Table 5. Vehicle arrival rates during each cycle

Cycle	Flow rate (veh/hr)	Flow rate (veh/sec)
1	900	.25
2	720	.20
3	540	.15

Step 2. Using Equation 13, calculate the approach capacity and compare it to the arrival flow rate for each cycle.

$$c = s \times \left(\frac{g}{C}\right) = 1900 \; veh/hr \times \left(\frac{40 \; sec}{100 \; sec}\right) = 760 \; veh/hr$$

The arrival flow rate is less than the capacity for the second and third cycle. But in the first cycle, the volume of 900 veh/hr exceeds the capacity of 760 veh/hr. Because volume exceeds capacity, Equation 31 cannot be applied to determine the delay. We must instead use the graphical approach of the cumulative vehicle diagram or queue accumulation diagram.

Step 3. Calculate the queue service time g_s for the first cycle using Equation 27.

$$g_s = \frac{vr}{s-v} = \frac{(.250 \ veh/sec)(60 \ sec)}{.528 \ veh/sec - .250 \ veh/sec} = 54.0 \ sec$$

This result confirms the finding in step 2 that the volume exceeds capacity. The queue would take 54.0 sec to clear, longer than the effective green time of 40 sec.

Step 4. Construct the cumulative vehicle diagram and queue accumulation polygon for the three cycles using the flow rate data given above.

Cumulative vehicle diagram: The arrival line (solid line, Figure 57) shows the cumulative number of vehicles that arrive from the beginning of the first cycle (t=0) to the end of the third cycle (t=300). The departure line (dashed line) shows the cumulative number of departures over time. The number of vehicles in queue at the end of each of the effective red and effective green periods are shown by the vertical lines.

Queue accumulation polygon. Figure 58 shows the variation in the queue length during the three cycles. The queue does not clear by the end of the first cycle, consistent with the calculations shown in steps 2 and 3. The residual queue at this point is 3.9 vehicles. Although the volume is less than the capacity in the second cycle, it is not sufficiently less to allow the queue to clear. There is still a residual queue of 2.8 vehicles at the end of cycle 2.

During the third cycle, the queue finally clears. But since there is a residual queue at the end of the second cycle, the numerator of Equation 27 (used to calculate the queue service time) must include both this residual queue and the queue that forms during effective red in the third cycle. The queue clears 31.2 seconds after the beginning of the effective green.

$$g_s = \frac{(residual \ queue) + vr}{s - v} = \frac{2.8 \ veh + (.15 \ veh/sec)(60 \ sec)}{(.528 \ veh/sec - .150 \ veh/sec)}$$

$$g_s = 31.2 \ sec$$

Step 5. Determine the total delay. The total delay can be calculated as the area under the curve in the cumulative vehicle diagram or queue accumulation polygon. Here we will use the queue accumulation polygon, divided into six separate polygons to facilitate the calculation of the area as shown in Figure 59.

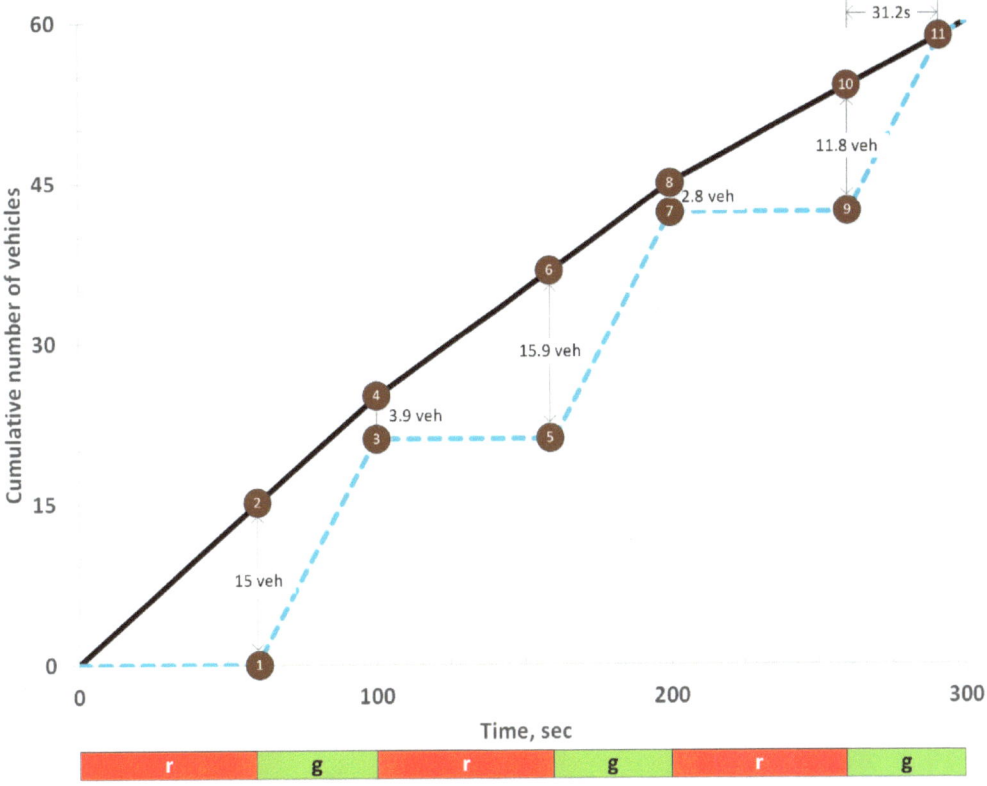

Figure 57. Cumulative vehicle diagram for Example 14

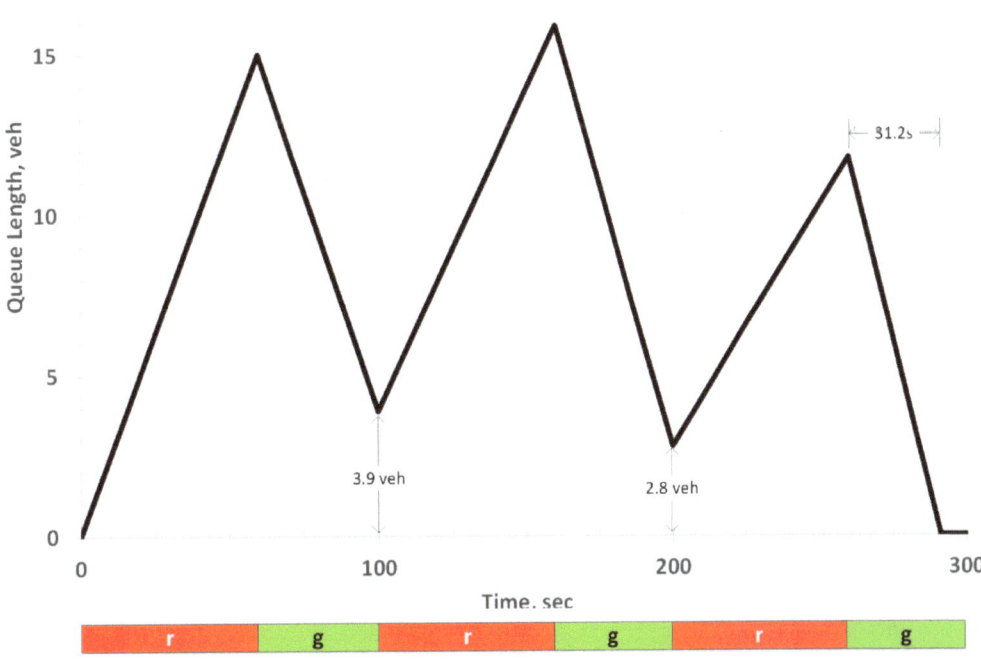

Figure 58. Queue accumulation polygon for Example 14

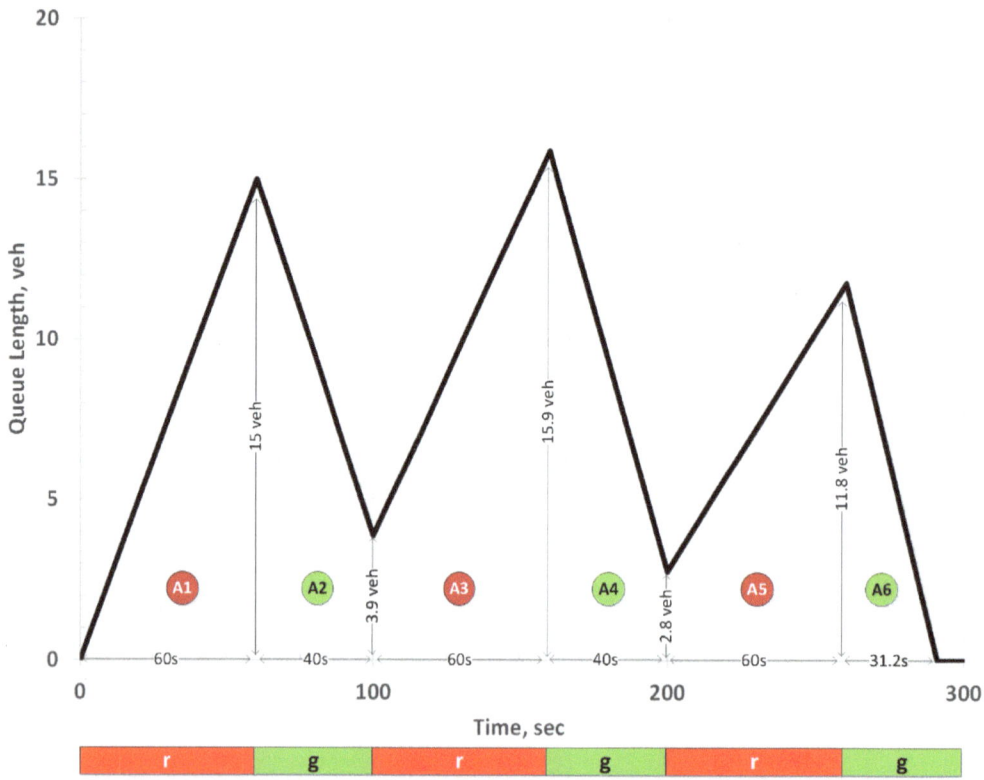

Figure 59. Constituent polygons from queue accumulation polygon

The calculations to determine the areas of each of the six constituent polygons are shown below.

Cycle 1:

$$A1 = \frac{15 \ veh}{2} \times (60 \ sec) = 450 \ veh - sec$$

$$A2 = \frac{3.9 \ veh + 15 \ veh}{2} \times (100 \ sec - 60 \ sec) = 378 \ veh - sec$$

Cycle 2:

$$A3 = \frac{15.9 \ veh + 3.9 \ veh}{2} \times (160 \ sec - 100 \ sec) = 594 \ veh - sec$$

$$A4 = \frac{2.8 \ veh + 15.9 \ veh}{2} \times (200 \ sec - 160 \ sec) = 374 \ veh - sec$$

Cycle 3:

$$A5 = \frac{11.8 \ veh + 2.8 \ veh}{2} \times (260 \ sec - 200 \ sec) = 438 \ veh - sec$$

$$A6 = \frac{11.8 \ veh}{2} \times (291.2 \ sec - 260 \ sec) = 184 \ veh - sec$$

7. Delay and Level of Service

The sum of these areas is the total delay for three cycles.

$$D_t = A_1 + A_2 + A_3 + A_4 + A_5 + A_6$$

$$D_t = 2418 \; veh - sec$$

Step 6. Determine the number of vehicles that arrive during the three cycles. Based on the arrival rates given in Table 5 and the length of the cycle (C = 100 sec), the total number of vehicles arriving during the three cycles is calculated below. Note that the number of vehicle arrivals can also be determined from the cumulative vehicle diagram in Figure 57.

Cycle 1:

$$Vehicle \; arrivals = (.25 \; veh/sec)(100 \; sec) = 25 \; veh$$

Cycle 2:

$$Vehicle \; arrivals = (.20 \; veh/sec)(100 \; sec) = 20 \; veh$$

Cycle 3:

$$Vehicle \; arrivals = (.15 \; veh/sec)(100 \; sec) = 15 \; veh$$

The total arrivals during three cycles:

$$Vehicle \; arrivals = 25 \; veh + 20 \; veh + 15 \; veh = 60 \; veh$$

Step 7. Calculate average delay.

$$d_{avg} = \frac{D_t}{number \; vehicles \; arrived} = \frac{2418 \; veh - sec}{60 \; veh} = 40.3 \; sec$$

7.2 Other Performance Measures for a Signalized Intersection

In addition to delay, other measures are used to estimate how well (or poorly) an intersection is operating. Two of these measures are the volume-to-capacity ratio and the back of queue. The volume-to-capacity ratio was discussed earlier in section 7 of this module.

The *back of queue* is the position of the vehicle stopped farthest from the stop line as a result of the red signal indication. It is useful for determining some of the geometric elements of the intersection, such as the length of a left turn bay. It is also useful in determining whether the operation of an upstream intersection will be affected by a spillback of the queue. The *back of queue size* (Q_{back}, the number of vehicles from the stop bar to the back of queue when the queue clears) is distinguished from the *maximum queue length Q_{max}*, the maximum number of vehicles in the queue, which occurs at the end of the effective red. These parameters are illustrated in Figure 60.

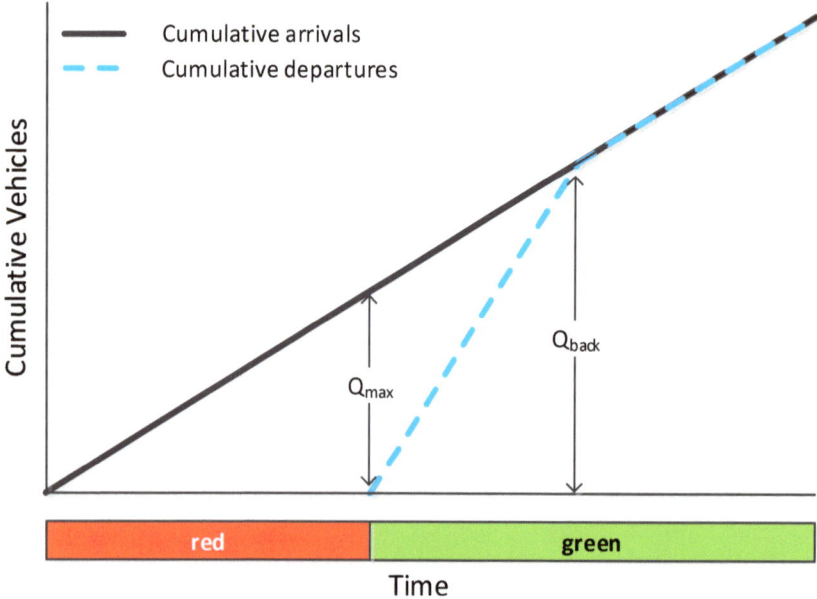

Figure 60. Maximum queue and back of queue size

The back of queue size is found by computing the product of the arrival rate and the sum of the effective red time and the queue service time, as given in Equation 33.

Equation 33

$$Q_{back} = v(r + g_s)$$

where

Q_{back} = back of queue size, veh,
v = arrival flow rate, veh/sec,
r = effective red time, sec,
g_s = queue service time, sec, and

Example 15. Back of Queue Size Calculation

The intersection shown in Figure 61 has an afternoon peak-hour volume of 250 veh/hr (.069 veh/sec) for the westbound left turn movement. The saturation flow rate is 1900 veh/hr (.528 veh/sec). The afternoon cycle length is 80 sec and the effective green for the left turn approach is 12 sec. Assume that vehicles maintain a 25 ft spacing while in queue (that is, the distance from the front of one vehicle to the front of the following vehicle is 25 ft). Calculate the back of queue size and determine whether the left turn bay of 125 feet is of sufficient length to accommodate this length of queue assuming that left turns are protected.

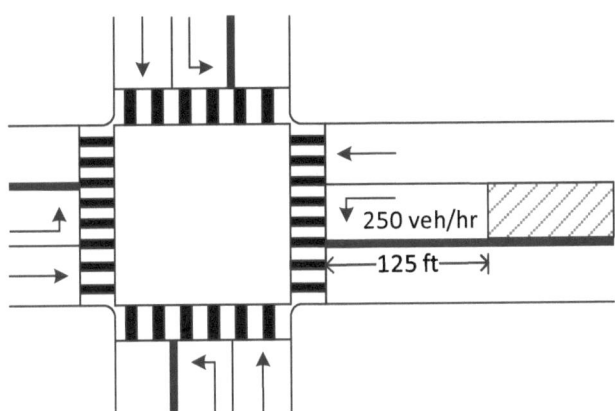

Figure 61. Intersection geometry and approach volume

Step 1. Calculate the queue service time using Equation 27.

$$g_s = \frac{vr}{s-v} = \frac{(.069 \; veh/sec)(68 \; sec)}{(.528 \; veh/sec - .069 \; veh/sec)} = 10.3 \; sec$$

Step 2. Calculate the back of queue size using Equation 33.

$$Q_{back} = v(r + g_s) = (.069 \; veh/sec)(68 \; sec + 10.3 \; sec) = 5.4 \; veh$$

Step 3. Calculate the distance upstream from the intersection stop bar reached by the back of queue.
Given the average vehicle spacing of 25 ft (and assuming 6 vehicles, closer to reality than the estimate of 5.4 vehicles), the distance upstream from the stop bar to the back of queue is:

$$distance = (6 \; veh)(25 \; ft/veh) = 150 \; ft$$

Step 4. Compare the distance reached by the back of queue to the storage available in the left turn bay.
The distance reached by the queue is 150 ft upstream of the stop bar. But the space available in the storage bay is 125 ft. There is not sufficient space available in the left turn bay to store the last (sixth) vehicle in the queue. This could result in a left turn vehicle blocking the westbound through movement, which could be both capacity problem and a safety problem.

7.3 Level of Service

The Highway Capacity Manual defines *quality of service* as how well a transportation facility operates from the perspective of the users of that facility. *Level of service* is a "quantitative stratification of a performance measure that represents the quality of service." For a signalized intersection, average delay is used as the performance measure. The HCM provides level of service ranges for a signalized intersection as shown in Table 6. These ranges can be applied to an intersection approach or to the entire intersection. It should be noted that if the volume-to-capacity ratio exceeds one, the level of service will be F, regardless of the estimated delay.

Table 6. Level of service ranges for a signalized intersection

Level of Service	Average control delay per vehicle (sec)
A	≤ 10
B	> 10 and ≤20
C	> 20 and ≤35
D	> 35 and ≤55
E	> 55 and ≤80
F	> 80

Example 16. Estimation Of Intersection Delay And Level Of Service

The demand and delay for four approaches of a signalized intersection are shown in Table 7. Determine the level of service for each approach and for the intersection as a whole.

Table 7. Traffic volume and delay for intersection approaches

Approach	Volume (veh/hr)	Volume (veh/sec)	Delay (sec)
Northbound	650	.181	25
Southbound	850	.236	18
Eastbound	200	.056	60
Westbound	300	.083	50

Step 1. Determine the level of service for each approach.
Based on the level of service ranges provided in Table 6, the level of service for each approach is shown in Table 8.

Table 8. Traffic volume and delay for intersection approaches

Approach	Volume (veh/hr)	Volume (veh/sec)	Delay (sec)	LOS
Northbound	650	.181	25	C
Southbound	850	.236	18	B
Eastbound	200	.056	60	E
Westbound	300	.083	50	D

Step 2. Determine the level of service for the intersection.
The volume and delay data given in Table 8 for each approach are used to determine the delay and level of service. The elements of the numerator for Equation 32 are calculated first:

$$d_{NB}v_{NB} = (25\ sec)(.181\ veh/sec) = 4.5\ veh$$

$$d_{SB}v_{SB} = (18\ sec)(.236\ veh/sec) = 4.2\ veh$$

$$d_{EB}v_{EB} = (60\ sec)(.056\ veh/sec) = 3.4\ veh$$

$$d_{EB}v_{EB} = (50\ sec)(.083\ veh/sec) = 4.2\ veh$$

7. Delay and Level of Service

Then, using Equation 32, the average delay for the intersection is:

$$d_{int} = \frac{\sum d_i v_i}{\sum v_i}$$

$$= \frac{(4.5 + 4.2 + 3.4 + 4.2)}{(.181 + .236 + .056 + .083)\ veh/sec} = 29.3\ sec$$

An average delay of 29.3 sec corresponds to an intersection level of service of C. This is acceptable. However, the level of service on the eastbound approach is E, nearly unacceptable. The westbound approach also experiences significant delay, at level of service D. The operation of the intersection may be improved if the signal timing were altered so that the minor approaches had more green time. The topic of signal timing is covered in the section 9.

7.4 Summary of Section 7

What You Should Know and Be Able to Do:
- Calculate uniform delay at a signalized intersection
- Describe and apply measures of effectiveness for a signalized intersection
- Describe and apply the level of service framework

Concepts You Should Understand:
- Concept 7.1: Queue service time

The queue service time is the ratio of the length of the queue at the beginning of the effective green time and the rate at which the queue clears after the beginning of the effective green.

- Concept 7.2: Uniform delay and the queue accumulation polygon

The total uniform delay for one movement during one cycle can be represented as the area of the queue accumulation polygon. The average uniform delay is the total uniform delay divided by the number of vehicles that arrive during the cycle.

8. CYCLE LENGTH AND SPLIT TIMES

> **Learning objectives**:
> - Describe the effect that cycle length has on intersection operation
> - Calculate the minimum cycle length
> - Relate the critical movement analysis to the determination of the split times

The length of the cycle has an important effect on the quality of service that motorists experience. Long cycle lengths increase delay while very short cycle lengths are inefficient. Also, the manner in which the cycle length is split between the phases has a significant effect on the quality of service experienced by motorists using the intersection. This section describes the methods used to determine the cycle length and the split times.

8.1 Cycle Length

The *cycle length* is the time required to serve one complete sequence of signal indications (green, yellow, and red clearance) for all intersection phases. The concept of cycle length can be represented using a ring-barrier diagram as shown in Figure 62. The length of the cycle is equal to the sum of the durations of the phases for ring 1 or ring 2. For a pretimed signal, the cycle length is constant.

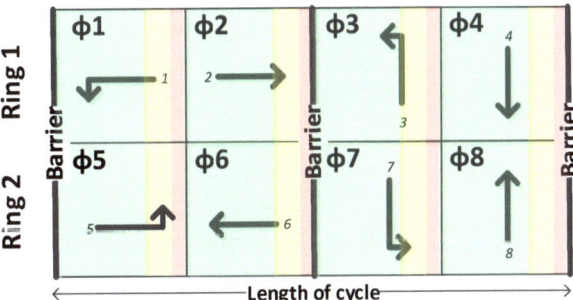

Figure 62. Cycle length representations

Cycle lengths that are too short can cause delays due to phase failures (when not all vehicles are served during a given phase). Very short cycle lengths are also inefficient since a large proportion of the cycle time is taken up with the lost time associated with the stopping of one phase and the starting of the next. But cycle lengths that are too long can cause excessive delay. The longer the cycle length, the longer the red time, and thus the longer the queues that build during red.

In practice, cycle lengths are generally kept as short as possible, typically between 60 and 90 sec. However, complex intersections with high volumes may have cycle lengths of 120 sec or more. Several methods are used in practice to determine the cycle length for an intersection, often using computer optimization models.

The method presented here is one suggested in the Highway Capacity Manual [4] and developed earlier by Pignataro [5]. It is based on the critical movement analysis method presented in section 6. The method produces a cycle length that is based on the minimum time required to serve all of the critical movements. The method is explained in the following paragraphs.

The flow ratio Y_i represents the minimum proportion of the hour required to serve a movement i. The critical flow ratio is the maximum flow ratio or maximum flow ratio sum for the critical movement(s) in each concurrency group (depending on whether the left turn phasing is protected or permitted). The time required during an hour to serve the critical movements from both concurrency groups is the product of the number of seconds in an hour and the sum of the critical flow ratios.

Equation 34

$$T_{crit} = 3600(Y_{EW-critical} + Y_{NS-critical})$$

where

T_{crit} = time required during an hour to serve all of the critical movements, sec,
$Y_{EW-critical}$ = the critical flow ratio for the EW concurrency group, and
$Y_{NS-critical}$ = the critical flow ratio for the NS concurrency group.

As described in section 5, a portion of the hour is "lost" as a result of the ending of one phase and the starting of another. If t_{Li} is the lost time for phase i and if each cycle has M critical phases, the *total lost time per cycle L* is given by:

Equation 35

$$L = \sum_{i=1}^{M} t_{Li}$$

Since the number of cycles during an hour is 3600/C (where C is the length of each cycle), the total lost time during an hour L_h is given by:

Equation 36

$$L_h = L\left(\frac{3600}{C}\right)$$

The constraint is that the sum of the times required to serve all of the critical movements (as given by Equation 34) and the total lost time per hour (as given by Equation 36) must be less than or equal to the number of seconds in an hour:

Equation 37

$$3600(Y_{EW-critical} + Y_{NS-critical}) + L\left(\frac{3600}{C}\right) \leq 3600$$

8. Cycle Length and Split Times

Changing Equation 37 to an equality and solving for *C*, Equation 38 shows the *minimum cycle length* required to serve the critical movements and account for the time that is lost due to the transition between phases.

Equation 38

$$C_{min} = \frac{L}{1 - (Y_{EW-critical} + Y_{NS-critical})}$$

where

C_{min} = minimum cycle length, sec,
L = total lost time per cycle, sec,
$Y_{EW-critical}$ = critical flow ratio for the east-west concurrency group, and
$Y_{NS-critical}$ = critical flow ratio for the north-south concurrency group.

Using Equation 38, Figure 63 shows a plot of the minimum cycle length as a function of the sum of the critical flow ratios for three values of lost time per cycle (8 sec, 12 sec, and 16 sec) for a range of critical flow ratio sums from 0.7 to 1.0. These lost times correspond to 2, 3, and 4 critical phases per cycle, respectively. While this method provides useful guidance in selecting the cycle length, it may, for low volumes, produce a cycle length value that is too low. In practice, a value between 60 sec to 90 sec is often used, even when the method suggests a value that is less than this range.

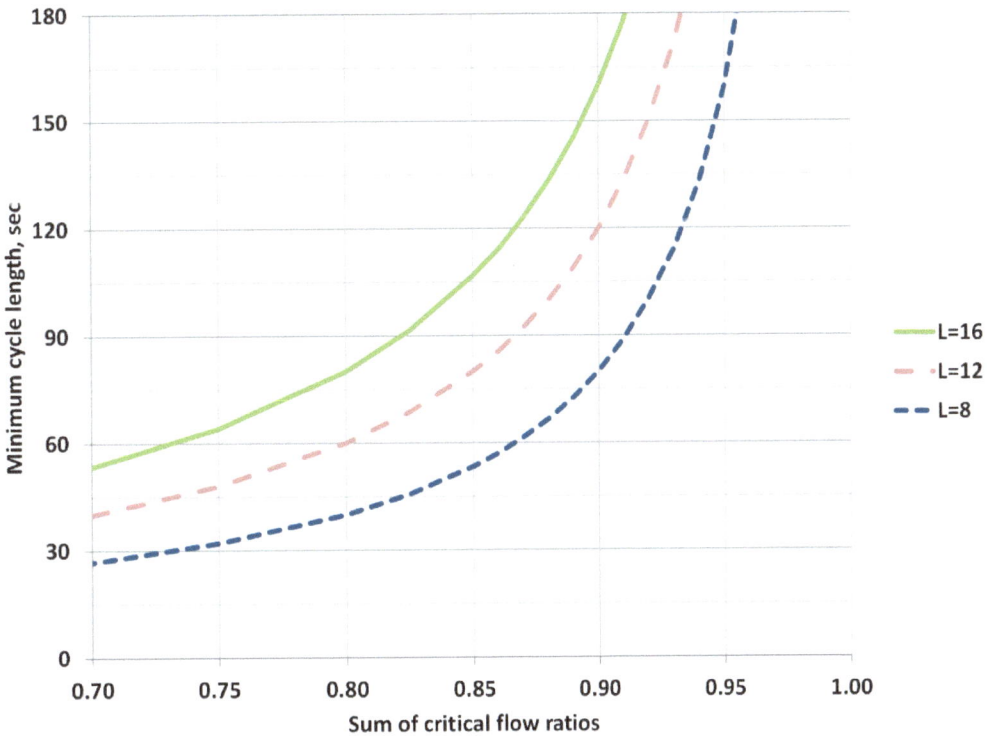

Figure 63. Cycle length as a function of the sum of the critical flow ratios

Example 17. Determining the Cycle Length

The flow ratios for an intersection have been computed and are listed in Table 9. The critical flow ratio sums for the east-west and north-south concurrency groups are .216 and .474, respectively. The total lost time per cycle is 16 seconds. What value would you recommend for the cycle length?

Table 9. Flow ratios for each movement

	East-West Concurrency Group		North-South Concurrency Group	
Ring 1	Y_1=.079	Y_2=.105	Y_3=.184	Y_4=.237
	Y_{EW1} = .184		Y_{NS1} = .421	
Ring 2	Y_5=.105	Y_6=.111	Y_7=.158	Y_8=.316
	Y_{EW2} = .216		Y_{NS2} = .474	

Using Equation 38,

$$C_{min} = \frac{L}{1 - (Y_{EW-critical} + Y_{NS-critical})} = \frac{16 \; sec}{1 - (.216 + .474)} = 51.6 \; sec$$

It is common in practice for the cycle length to be rounded up to the nearest 5 sec or in this example to 55 sec. But since this value is less than the minimum value suggested in the discussion above, a cycle length of 60 sec would be used here.

8.2 Split Times

The determination of green times for each phase is a process of splitting the cycle based on the respective flow ratios for each critical movement. The term *split* is often used to describe the result of this process. The *split time* for each phase consists of the displayed green time, the yellow time, and the red clearance time.

Equation 39

$$S_i = G + Y + RC$$

where

S_i = split time for phase i, sec, and

G, Y, and RC = displayed green time, yellow time, and red clearance time for phase i, sec.

When the critical movements have been determined (as described in section 6), the flow ratio Y_i for each critical movement is used to determine the proportion of the cycle required to serve that movement. This proportion is called the *split proportion* and is given in Equation 40.

8. Cycle Length and Split Times

Equation 40

$$SP_i = \frac{Y_i}{Y_{EW-critical} + Y_{NS-critical}}$$

where

SP_i = split proportion to serve critical movement i, sec,

Y_i = flow ratio for critical movement i,

$Y_{EW-critical}$ = critical flow ratio for the east-west concurrency group, and

$Y_{NS-critical}$ = critical flow ratio for the north-south concurrency group.

The split time S_i for phase i is equal to the product of the split proportion for phase i and the cycle length, and is given by Equation 41.

Equation 41

$$S_i = (SP_i)\,(C)$$

It is important to note that the minimum displayed green time is almost never less than 5 sec. So, if the process described above results in such a value, the displayed green time is set to 5 sec, and the remainder of the cycle time is reallocated (re-split) to the other critical phases in proportion to their flow ratios.

Example 18. Calculating Split Times

The flow ratios for each movement at a signalized intersection have been calculated and are summarized in Table 10. The cycle length was determined to be 60 sec, as per Example 16. What are the split times for the critical phases?

Table 10. Flow ratios for each movement

	East-West Concurrency Group		North-South Concurrency Group	
Ring 1	Y_1=.079	Y_2=.105	Y_3=.184	Y_4=.237
Ring 2	Y_5=.105	Y_6=.111	Y_7-.158	Y_8=.316

The split proportion for each of the critical phases (5, 6, 7, and 8) is computed using Equation 40. The split time is computed using Equation 41. The results are summarized in Table 11.

Table 11. Split times for critical phases

	East-West Concurrency Group		North-South Concurrency Group		Total
Critical phase	5	6	7	8	
Flow ratio	.105	.111	.158	.316	.690
Split proportion	.152	.161	.229	.458	1.000
Split time (sec)	9.1	9.7	13.7	27.5	60

This example illustrates an important point in the determination of the green times. In most traffic signal systems, the minimum green time displayed to the user is 5 sec. Let's assume for this example that the sum of the yellow and red clearance times is equal to 5 sec. For phases 5 and 6, the split time minus the yellow and red clearance times result in a green display of less than 5 sec (4.1 and 4.7 sec, respectively).

To resolve this problem, the split times for phases 5 and 6 would each be set to 10 sec. The remaining time (60 sec - 10 sec – 10 sec) of 40 sec is allocated to phases 7 and 8, based on their relative flow ratio proportions. To illustrate this point, the relative flow ratio proportion for phase 7 is calculated as:

$$\frac{.158}{.158 + .316} = .333$$

The relative flow ratio proportion for phase 8 is calculated in a similar manner.

$$\frac{.316}{.158 + .316} = .667$$

These results are summarized in Table 12.

Table 12. Modified green times for critical phases

	East-West Concurrency Group		North-South Concurrency Group		Total
Critical phase	5	6	7	8	
Flow ratio	.105	.111	.158	.316	.690
Initial split proportion	.152	.161	.229	.458	1.000
Initial split time (sec)	9.1	9.7	13.7	27.5	60
Minimum split times (sec)	10.0	10.0			
Relative flow ratio proportions			.333	.667	1.000
Modified split times (sec)	10.0	10.0	13.3	26.7	60

While the split times for the non-critical phases can be adjusted to provide for efficiency, here they are set equal to the split times for the respective critical phases in the other ring. For example, the split time for phase 1 (non-critical phase) would be set equal to the split time for phase 5 (critical phase). These split times are shown in Table 13.

Table 13. Split times for critical and non-critical phases

Critical phase	5	6	7	8
Corresponding non-critical phase	1	2	3	4
Split time (sec)	10.0	10.0	13.3	26.7

8.3 Summary of Section 8

What You Should Know and Be Able to Do:

- Describe the effect that cycle length has on intersection operation
- Calculate the minimum cycle length
- Relate the critical movement analysis to the determination of the split times

8. Cycle Length and Split Times

Concepts You Should Understand:

- Concept 8.1: Minimum cycle length

The minimum cycle length is the time required to serve the critical movements as well as to account for the time that is lost due to the transition between phases.

- Concept 8.2: Flow ratio, split proportions, and split times

The flow ratios for the critical movements are used to determine the split proportions and the split times for each phase.

9. SIGNAL TIMING

Learning objectives:
- Determine appropriate intersection phasing
- Apply understanding of the ring barrier diagram and critical movement analysis to estimate cycle length and allocate green times
- Calculate change and clearance intervals
- Compare pedestrian crossing time requirements and split times

Assuming the decision to install a traffic signal at an intersection has been made, a phasing and timing plan must then be developed. This section integrates the concepts developed in the earlier sections of this module to describe a process to develop a phasing and timing plan for an isolated, pretimed traffic signal. This process (shown in Figure 64) includes the following seven steps.
- Step 1. Select signal phasing
- Step 2. Compute flow ratios
- Step 3. Determine critical movements
- Step 4. Determine yellow and red clearance times
- Step 5. Determine cycle length
- Step 6. Determine split times
- Step 7. Check pedestrian crossing times

Once the phasing, the cycle length, the green times, the yellow times, and the red clearance times are determined, the intersection performance can be evaluated, as shown in step 8.

The development of a phasing and timing plan can be complex, particularly if the intersection has multilane approaches and moderate to heavy left turning movements. The reader is encouraged to review other references [2, 4] for information on actuated control and signal coordination. As intersections become more complex, the phasing and timing plans needed to control them build on the fundamental principles presented in this module.

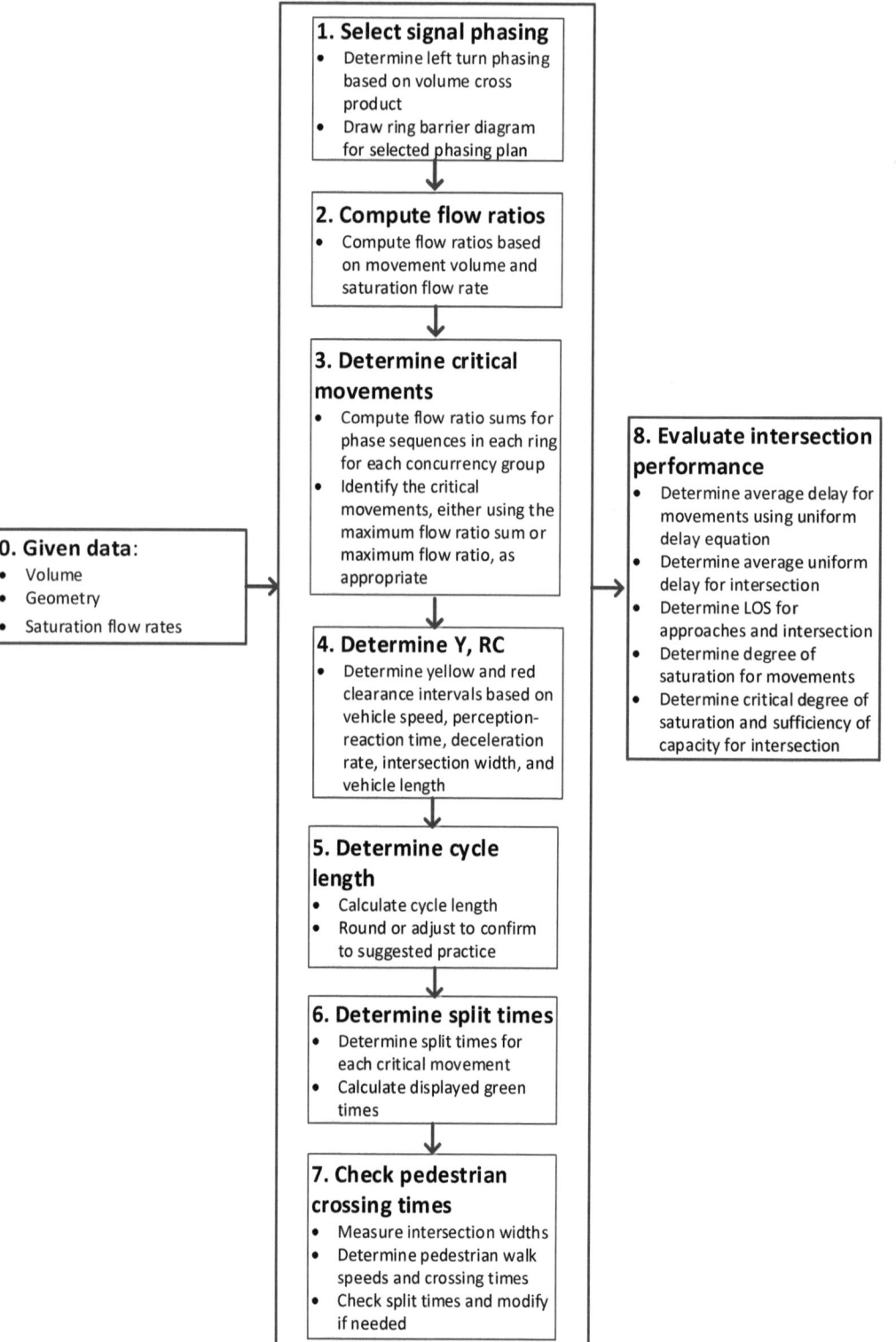

Figure 64. Signal timing design and intersection evaluation process

9. Signal Timing

To illustrate the development of the phasing and timing plan, the intersection shown in Figure 65 will be used throughout this section in a set of example calculations covering the eight steps listed above. The through movements and protected left turn movements have a saturation flow rate of 1900 veh/hr/lane while permitted left turn movements have a saturation flow rate of 450 veh/hr/lane. The speed limit on both streets is 35 mi/hr (51.3 ft/sec). The volumes and intersection widths are shown in Figure 65.

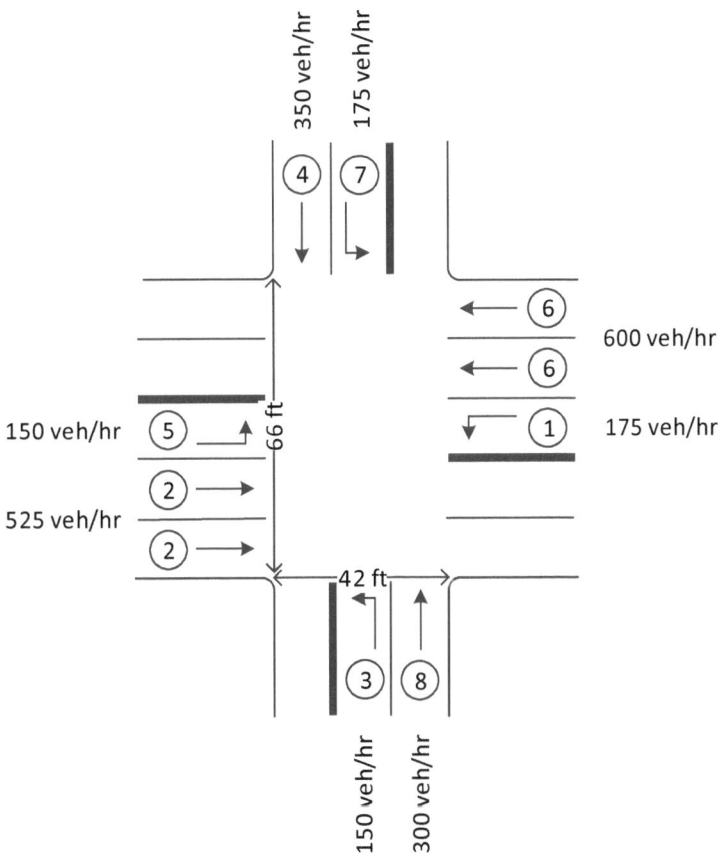

Figure 65. Example intersection

9.1 Step 1: Select Signal Phasing

The signal phasing plan identifies which movements are present, the phases that will control the movements, and the order in which the phases are sequenced. The phasing plan is described in terms of a ring barrier diagram, a concept introduced in section 3. The primary decision to be made is whether left turns will be protected or permitted. While permitted left turn phasing is usually more efficient in terms of minimizing lost time during a cycle, protected left turn phasing should be considered when the left turn volume cross product exceeds the criteria listed in section 3.

Example 19-Step 1: Select Signal Phasing

- Step 1a: Identify the left turn-opposing through movement volume pairs. Find the cross-product of the volumes of these movement pairs, as shown in Figure 66. For the east-west concurrency group, the reference value is 90,000 because the left turns are opposed by two through lanes. For the north-south concurrency group, the reference value is 50,000 because the left turns are opposed by one through lane. Since the volume cross-products exceed the reference values, protected left turn phasing will be used. These results are summarized in Figure 66.

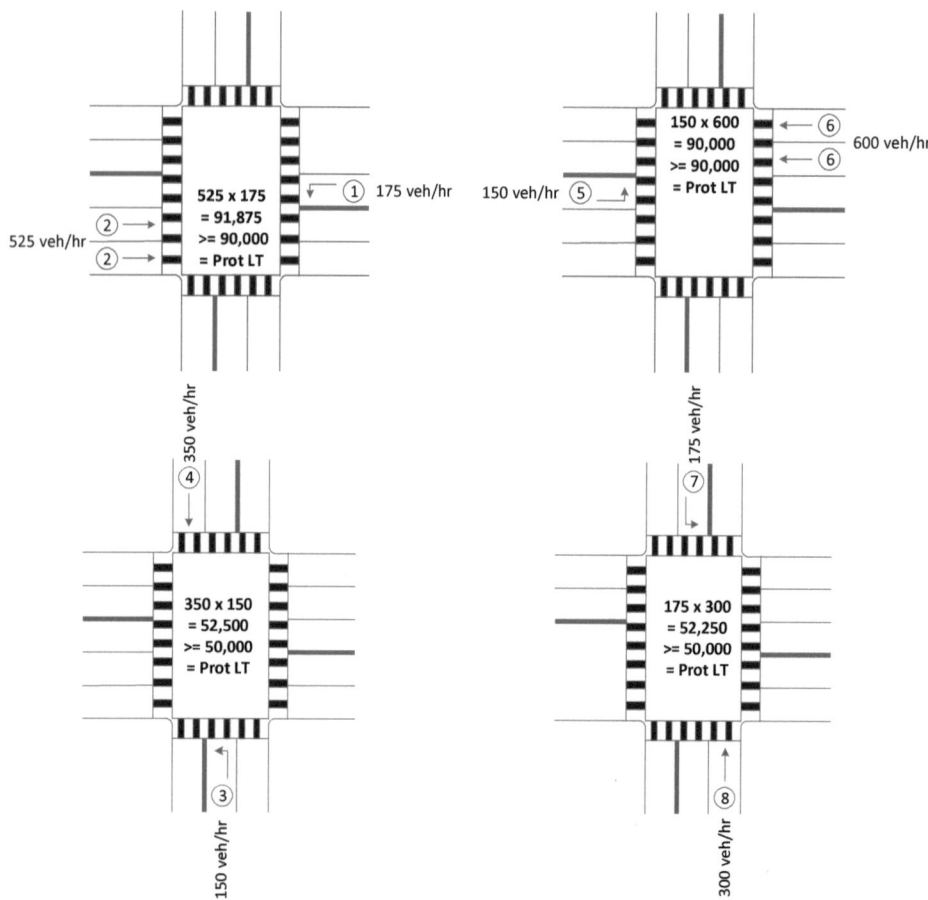

Figure 66. Left turn protection calculations

- Step 1b: Determine phasing plan. Because all left turns should be protected, the standard 8-phase two-ring control should be used. This phasing plan is represented with the ring barrier diagram shown in Figure 67.

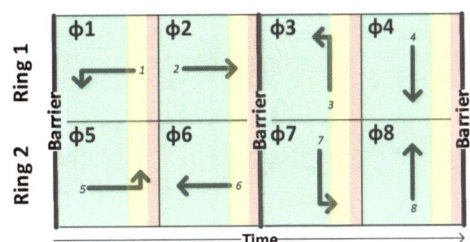

Figure 67. Phasing plan

9.2 Step 2: Compute Flow Ratios

The flow ratios are computed as described in section 7.

Example 19-Step 2: Compute flow ratios for each movement present at the intersection.

The flow ratios for each movement are calculated and shown in Figure 68.

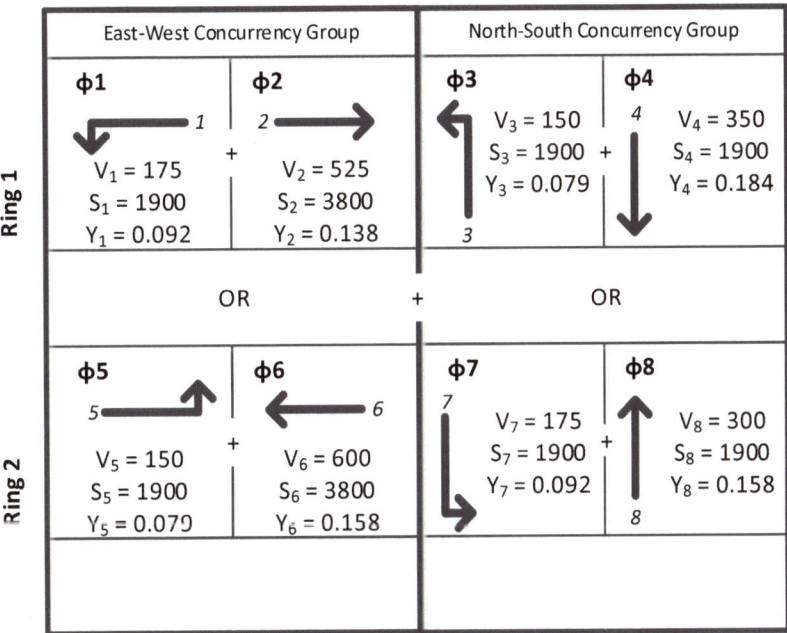

Figure 68. Flow ratios by ring and concurrency group

9.3 Step 3: Determine Critical Movements

The critical movements are determined using the procedure described in section 7. The critical movement analysis determines which sequence of phases requires the longest green time.

Example 19-Step 3: Determine the Critical Movements

Step 3a: Determine the flow ratio sums for the phase sequences in each ring for each concurrency group (for the case of protected left turns only).

The flow ratio sums for each ring within each concurrency group are shown in Figure 69.

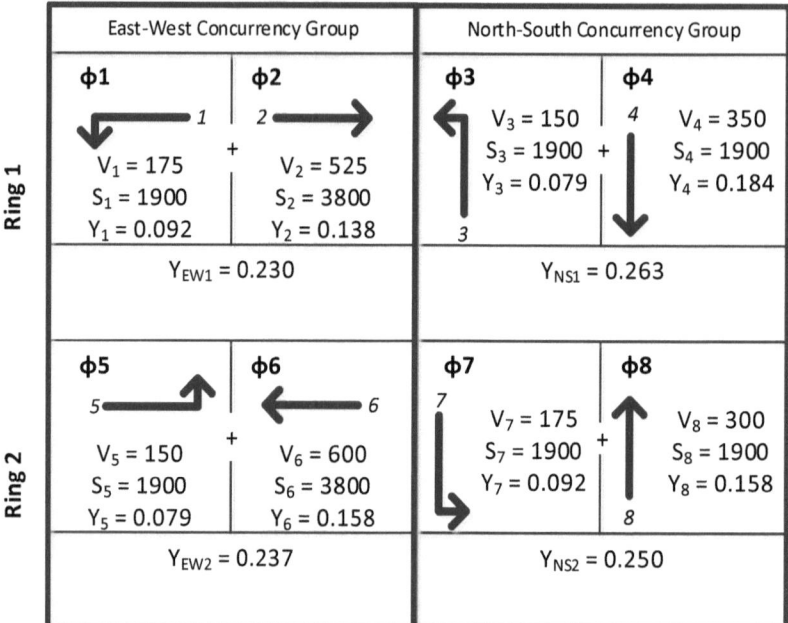

Figure 69. Flow ratio sums by ring and concurrency group

Step 3b: Within each concurrency group, identify the movements with the maximum flow ratio sum (for protected left turns) or the movement with the maximum flow ratio (for permitted left turns). These movements are the critical movements for each concurrency group.

Since these are protected left turns, we identify the movements with the maximum flow ratio sum within each concurrency group (see Figure 70). For the east-west concurrency group, the movements served in ring 2 (movements 5 and 6) have the highest flow ratio sum (.237), as compared to the movements served in ring 1 (.230). For the north-south concurrency group, the movements served in ring 1 have the highest flow ratio sum (.263).

$$Y_{EW-critical} = Max(Y_{EW1}, Y_{EW2}) = Max(.230, .237) = .237$$

$$Y_{NS-critical} = Max(Y_{NS1}, Y_{SW2}) = Max(.263, .250) = .263$$

9. Signal Timing

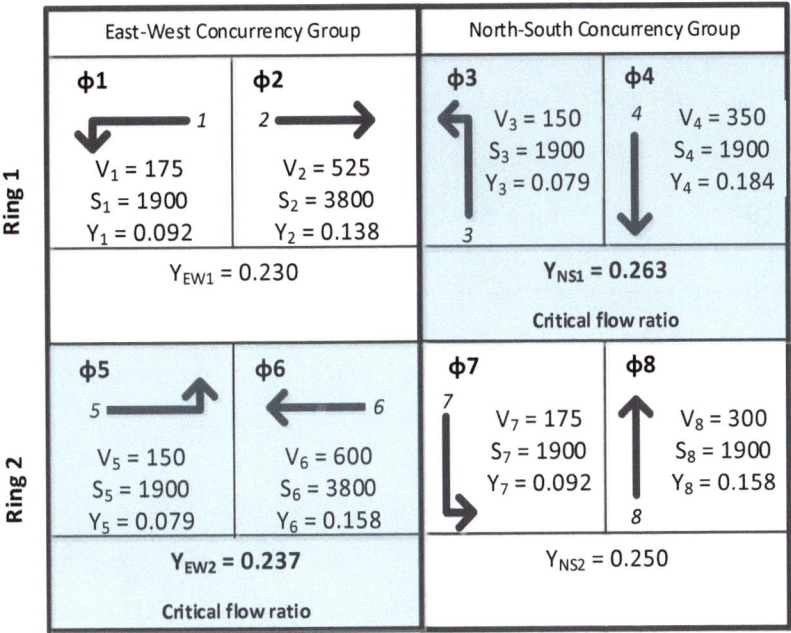

Figure 70. Critical flow ratio by concurrency group

9.4 Step 4: Determine Yellow and Red Clearance Times

The yellow and red clearance times are calculated using the method presented in section 5.

Example 19-Step 4: Calculate the Change and Clearance Intervals

The approach speed is given as 35 mi/hr (51.3 ft/sec) and the comfortable deceleration rate is assumed to be 10 ft/sec². The intersection widths are 42 ft and 66 ft (see Figure 65). The average vehicle length is assumed to be 20 ft and the perception reaction time is given as 1.0 sec..

Step 4a. Yellow time
Equation 4 is used to calculate the yellow change interval.

$$Y = \delta + \frac{v}{2a} = 1.0 \ sec + \frac{51.3 \ ft/sec}{2(10 \ ft/sec^2)} = 3.6 \ sec$$

Step 4b: Red clearance time
The red clearance is calculated using Equation 5.

East-west approaches:

$$RC = \frac{w + L}{v} = \frac{42 \ ft + 20 \ ft}{51.3 \ ft/sec} = 1.2 \ sec$$

North-south approaches:

$$RC = \frac{w + L}{v} = \frac{66 \ ft + 20 \ ft}{51.3 \ ft/sec} = 1.7 \ sec$$

9.5 Step 5: Determine Cycle Length

The minimum cycle length is calculated using Equation 38 as described in section 9, using the lost time per cycle and the critical flow ratio for each concurrency group.

Example 19-Step 5: Determine the Cycle Length

From step 2 of this example, the critical flow ratios for the east-west and north-south concurrency groups are .237 and .263, respectively. The sum for the intersection is .500. Since there is a sequence of four critical phases, the total lost time is 16 sec/cycle assuming 4 sec/phase. The minimum cycle length can be calculated using Equation 38:

$$C_{min} = \frac{L}{1 - (Y_{EW-critical} + Y_{NS-critical})} = \frac{16 \ sec}{1 - (.237 + .263)} = 32 \ sec$$

This computed cycle length is reset to 60 sec. Generally, if the computed cycle length is than 60 sec, we would use this lower bound as the final selection of the cycle length.

9.6 Step 6: Determine Split Times

The split times and the displayed green times are calculated using the methods from section 9.

Example 19-Step 6: Determine the Split Times and the Effective Green Times for Each Phase.

Step 6a: The split proportion for each of the critical phases is calculated as the flow ratio for the phase divided by the sum of the critical flow ratios. For phase 5, the split proportion is calculated as follows:

$$SP_5 = \frac{0.079}{0.500} = .158$$

Step 6b: The split time required for each critical phase is calculated using Equation 41. Using the cycle length determined in step 3 (C = 60 sec), the split time for phase 5 is computed as

$$S_5 = (0.158)(60 \ sec) = 9.5 \ sec$$

The split times for the other critical movements are calculated similarly and shown in Table 14.

9. Signal Timing

Table 14. Split times for critical phases

	East-West Concurrency Group		North-South Concurrency Group		Total
Critical phase	5	6	3	4	
Flow ratio	0.079	0.158	0.079	0.184	.500
Split proportion	0.158	0.316	0.158	0.368	1.000
Split time (sec)	9.5	18.9	9.5	22.1	60

Step 6c: Since the split times calculated for phases 3 and 5 are less than the minimum split time (minimum displayed green of 5 seconds plus the yellow and red clearance times calculated in step 4), they are reset to 10.4 and 9.9 seconds, respectively. The split times for phases 4 and 6 are recalculated based on their relative flow ratio proportions and the remaining unallocated split time of 39.7 sec. The results are shown in Table 15.

Table 15. Adjusted split times for critical phases

	East-West Concurrency Group		North-South Concurrency Group		Total
Critical phase	5	6	3	4	
Flow ratio	0.079	0.158	0.079	0.184	.500
Split proportion	0.158	0.316	0.158	0.368	1.000
Split time (sec)	9.5	18.9	9.5	22.1	60
Minimum split time (sec)	9.9		10.4		
Relative flow ratio proportion		0.462		0.538	1.000
Adjusted split time (sec)	9.9	18.3	10.4	21.4	60

Step 6d: Green times

The displayed green time is equal to the split time minus the yellow and red clearance times. Using the split times from step 6c, and the yellow and red clearance times calculated in steps 4a and 4b, the resulting green times are shown in Table 16. As a simplification, the split times for the non-critical phases are set equal to the split times for their respective compatible phases in the other ring.

Table 16. Displayed green times

	East-West Concurrency Group		North-South Concurrency Group	
Phase	**1/5**	**2/6**	**3/7**	4/8
Split time (sec)	9.9	18.3	10.4	21.4
Yellow time (sec)	3.6	3.6	3.6	3.6
Red clearance time (sec)	1.3	1.3	1.8	1.8
Displayed green time (sec)	5.0	13.4	5.0	16.0

9.7 Step 7: Check Pedestrian Crossing Times

While pedestrian timing has not been explicitly covered in this module, it is important to check that the displayed green time in each vehicle phase is equal to or greater than the sum of the two pedestrian timing intervals, the WALK interval and the flashing DON'T WALK, for the compatible pedestrian phase. The

WALK display is a minimum of 4 seconds, though it may be longer when pedestrian flows are heavy and cycle lengths are long. The flashing DON'T WALK interval (which typically occurs only during the green interval) is equal to the *pedestrian clearance time* (PCT) minus the yellow and red clearance times, as shown in Equation 42.

Equation 42

$$FDW = PCT - Y - RC$$

where
   FDW = flashing DON'T WALK interval, sec
   PCT = pedestrian clearance time, sec
   Y = yellow time, sec, and
   RC = red clearance time, sec.

The pedestrian clearance time is the crossing time of the intersection based on the width of the intersection (pedestrian crossing distance) and an assumed pedestrian walking speed. The MUTCD [1] suggests a walking speed of 3.5 ft/sec.

Equation 43

$$PCT = \frac{w}{v_p}$$

where
   w = pedestrian crossing distance (usually intersection width), ft, and
   v_p = pedestrian walking speed, ft/sec.

The duration of the vehicle phase (the sum of the green, yellow, and red clearance displays) must be greater than or equal to the sum of the WALK interval and the pedestrian clearance time (PCT).

Equation 44

$$G + Y + RC \geq W + PCT$$

where
   G = displayed green, sec,
   Y = yellow time, sec
   RC = red clearance time, sec,
   W = pedestrian start up or WALK time, sec, and
   PCT = pedestrian clearance time, sec.

Or, based on the relationship defined in Equation 42, Equation 44 can be simplified to

Equation 45

$$G \geq W + FDW$$

with all variables as defined above.

The relationships between the pedestrian intervals and the vehicle signal displays are illustrated in Figure 71.

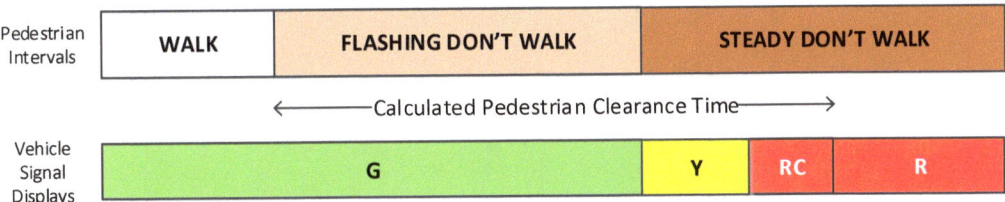

Figure 71. Relationship between pedestrian intervals and vehicle signal displays (Adapted from [2])

Example 19-Step 7: Check Pedestrian Walking Times

The results of the pedestrian walking time checks are shown in Table 17. The PCT for each crosswalk is computed based on a walking speed of 3.5 ft/sec and the curb-to-curb crossing distance (w). The sum of W and FDW is compared with the displayed green time (G) from Table 16 for each of the four intersection approaches. For phases 2 and 6, the sum of W and FDW is 11.2 sec while G is 13.4 sec. Thus, the duration of the green display is sufficient to accommodate these pedestrian timing intervals. For phases 4 and 8, however, the sum of W and FDW is 17.6 sec while G is 16.2 sec. Thus, the displayed green time is not adequate for phases 4 and 8, and must be increased to 17.6 sec.

Table 17. Pedestrian walking time check

Approach (Phase)	Concurrent Crosswalk	w (ft)	PCT (sec)	W (sec)	FDW (sec)	WALK + FDW (sec)	G (sec)	Green Time Adequate?
EB/WB (2/6)	N/S	42	12.0	4.0	7.2	11.2	13.4	Yes
NB/SB (4,8)	E/W	66	18.9	4.0	13.6	17.6	16.2	No

9.8 Final Signal Timing Design

How do we proceed with this design? We have the following constraints from the calculations in steps 6 and 7.

- The left turn phases (1, 3, 5, and 7) are at their minimum displayed green times of 5 sec.
- Two of the through phases (4 and 8) are fixed by the required pedestrian times of 17.6 sec.

Table 18 shows the constraints for the displayed green times as well as the yellow and red clearance times calculated previously. These results show that a cycle length of 61.4 sec is required based on the constraints listed above for the displayed green times. Following the practice described in section 8, the cycle length will be rounded up to 65 sec.

Table 18. Required display times

Phase	1/5	2/6	3/7	4/8
G (sec)	5.0	13.6	5.0	17.6
Y (sec)	3.6	3.6	3.6	3.6
RC (sec)	1.2	1.2	1.7	1.7
Split time (sec)	9.8	18.4	10.3	22.9
Total (sec)	61.4			

The split times are now adjusted based on the ratio of the rounded cycle length (65 sec) to the calculated cycle length (61.4 sec). For example, the split time shown in Table 18 for phases 1 and 5 is increased from 9.8 seconds to the value of 10.4 seconds shown in Table 19.

$$\left(\frac{65 \ sec}{61.4 \ sec}\right)(9.8 \ \text{sec}) = 10.4 \ sec$$

The final signal timing plan is summarized in Table 19. The plan includes the elements of a phasing and timing plan for a signalized intersection with pretimed control.

Table 19. Signal timing summary

	Phase or Movement Data							
	1	2	3	4	5	6	7	8
Left turn phasing	Prot		Prot		Prot		Prot	
Critical movement			X	X	X	X		
Cycle length, sec	65							
Split times, sec	10.4	19.5	10.9	24.2	10.4	19.5	10.9	24.2
Displayed green time, sec	5.6	14.7	5.7	19.0	5.6	14.7	5.7	19.0
Yellow time, sec	3.6	3.6	3.6	3.6	3.6	3.6	3.6	3.6
Red time, sec	1.2	1.2	1.7	1.7	1.2	1.2	1.7	1.7

9.9 Step 8: Evaluate Intersection Performance

We can now determine the performance of the intersection with this signal timing plan, including the critical volume-to-capacity ratio, the average delay, and the back of queue size. The results of the intersection performance evaluation are shown in Table 20 with three example calculations given below.

The critical volume-to-capacity ratio for the intersection is calculated using Equation 25:

$$X_c = \frac{(Y_{EW-critical} + Y_{NS-critical})(C)}{C - L} = \frac{(.237 + .263)(65)}{65 - 16} = 0.66$$

9. Signal Timing

The average delay is computed using Equation 31. The example below shows the calculation of the average delay for phase 4.

$$d_{avg} = (0.5r)\left[\frac{(1 - g/C)}{(1 - v/s)}\right] = (0.5)(44.8)\left[\frac{(1 - .311)}{(1 - .184)}\right] = 18.9 \; sec$$

The back of queue size is computed using Equation 33. The example below shows the calculation of the back of queue size for movement 4. The queue service time is computed using Equation 27.

$$g_s = \frac{vr}{s - v} = \frac{(350/3600)(44.8)}{(1900/3600) - (350/3600)} = 10.1 \; sec$$

$$Q_{back} = v(r + g_s) = (350/3600)(44.8 + 10.1) = 5.3 \; veh$$

Table 20. Intersection performance evaluation

	Phase or Movement Data							
	1	2	3	4	5	6	7	8
Volume, veh/hr	175	525	150	350	150	600	175	300
Saturation flow rate, veh/hr	1900	3800	1900	1900	1900	3800	1900	1900
Flow ratio	.092	.138	.079	.184	.079	.158	.092	.158
Effective green, sec	6.4	15.5	6.9	20.2	6.4	15.5	6.9	20.2
Effective red, sec	58.6	49.5	58.1	44.8	58.6	49.5	58.1	44.8
Average delay, sec/veh	29.1	21.9	28.2	18.9	28.7	22.4	28.6	18.3
Capacity	186	906	202	591	186	906	202	591
Volume-to-capacity ratio	0.94	0.58	0.74	0.59	0.80	0.66	0.87	0.51
Back of queue size, veh	3.1	8.4	2.6	5.3	2.7	9.8	3.1	4.4
Level of service	C	C	C	B	C	C	C	B
Average delay, sec/veh	23.0							
Level of service	C							
Critical v/c ratio	0.663							

The ring barrier diagram showing the phasing plan as well as the timings for each phase is shown in Figure 72.

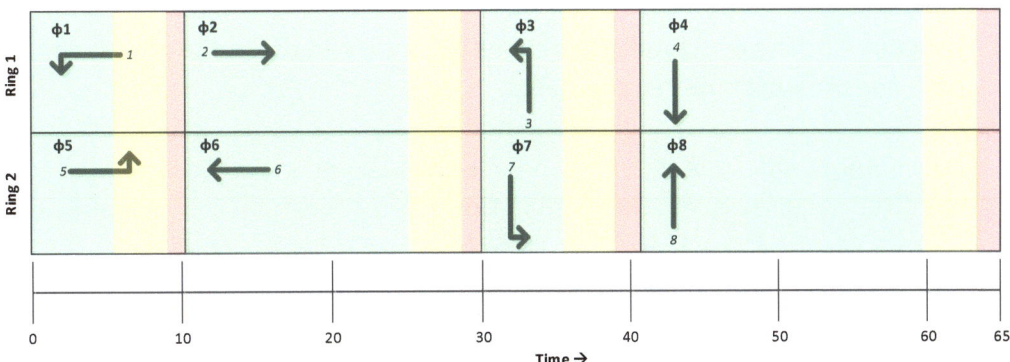

Figure 72. Ring barrier diagram and timings for final design

Overall, the intersection operates at less than capacity and, with an average delay of 23.0 seconds, at level of service C. But this intersection level view hides one potential operational problem. Movement 1 operates just below capacity, with a volume-to-capacity ratio equal to 0.94. Shifting the green times from the through movements to this left turn movement is one option to improve its operation without significantly degrading the operation of the other movements.

9.10 Summary of Section 9

What You Should Know and Be Able to Do:

- Determine appropriate intersection phasing
- Apply understanding of the ring barrier diagram and critical movement analysis to estimate cycle length and allocate green times
- Calculate change and clearance intervals
- Compare pedestrian crossing time requirements and split times

Concepts You Should Understand:

- Concept 9.1: Left turn phasing depends on the opposing left turn and through movement volumes. A standard from the Highway Capacity Manual is used to make a determination whether protected or permitted left turn phasing should be used.

- Concept 9.2: Critical flow ratios determine the amount of green time required for each phase.

- Concept 9.3: Yellow and red clearance times depend on the average speed of vehicles approaching the intersection as well as the width of the intersection.

- Concept 9.4: Split times depend on the relative flow ratios for each critical movement.

- Concept 9.5: Minimum green time is often a constraint in determining the displayed green time, despite the split times computed using the relative flow ratios.

- Concept 9.6: The displayed green times must exceed the pedestrian start up time and crossing time requirements.

- Concept 9.7: Adjustments to the initial calculated values of displayed green are often required to meet minimum times and pedestrian requirements.

- Concept 9.8: Performance measures such as delay help to identify potential operational problems that need attention and may suggest changes to the signal timing plan.

GLOSSARY

Table 21. Variables used in the module

Variable	Description	Units
a	Comfortable deceleration rate	ft/sec^2
C	Cycle length	sec
C_{min}	Minimum cycle length	sec
c	Capacity	veh/hr
D_t	Total uniform delay	veh-sec
d_{avg}	Average uniform delay per vehicle	sec
d_i	Average uniform delay for approach i	sec
d_i	Delay experienced by vehicle i	sec
d_{int}	Average uniform intersection delay	sec
G	Displayed green time	sec
g	Effective green time	sec
g_s	Queue service time	sec
h_a	Arrival headway	sec
h_s	Saturation headway	sec
L	Lost time per cycle	sec
L_h	Total lost time during an hour	sec
L_v	Vehicle length	ft
M	Number of critical phases per cycle	
PCI	Pedestrian clearance interval	sec
Q	Length of queue	vehicles
Q_{back}	Back of queue size	vehicles
Q_{max}	Maximum queue length	vehicles
r	Effective red time	sec
R	Displayed red time	sec
RC	Red clearance time	sec
s	Saturation flow rate	veh/hr
S_i	Split time for phase i	sec
SP_i	Split proportion for phase i	
T_{crit}	Time required during an hour to serve all critical movements	sec
t_i	Lost time for vehicle i	sec
t_L	Total lost time for a phase	sec
t_{sl}	Startup lost time	sec
t_{cl}	Clearance lost time	sec
v	Arrival flow rate or volume	veh/hr
v	Vehicle speed (velocity)	mi/hr or ft/sec
w	Intersection width	ft
$WALK$	Displayed WALK time or pedestrian startup time	sec
X	Volume-to-capacity ratio	
X_c	Critical volume-to-capacity ratio	
x_c	Clearing distance	ft
x_s	Stopping distance or distance from choice point to intersection stop bar	ft
Y	Yellow time	sec
Y	Flow ratio	
Y_i	Flow ratio for movement i	
$Y_{xx-crit}$	Critical flow ratio for concurrency group xx	
δ	Perception-reaction time	sec

Table 22. Terms used in the module

Term	Definition
Actuated control	A type of signal control in which the phase times depend on the traffic volume.
Arrival headway	The headway between vehicles arriving at a signalized intersection.
Average uniform delay	The average delay per vehicle experienced by vehicles during a cycle when the arrival flow is uniform.
Back of queue	"The maximum backward extent of queued vehicles during a typical cycle, as measured from the stop line to the last queued vehicle."[HCM 2010]
Back of queue size	The number of vehicles from the stop bar to the back of queue.
Capacity	"The maximum sustainable hourly flow rate at which persons or vehicles reasonably can be expected to traverse a point or a uniform section of a lane or roadway during a given time period under prevailing roadway, environmental, traffic, and control conditions." [HCM 2010]
Choice point	The closest point upstream of the stop bar at which the driver will be able to safely stop at the onset of yellow should he or she choose to do so; it is also the farthest point upstream of the stop bar at the onset of yellow at which the driver can safely and completely clear the intersection, if the driver chooses to do so.
Clearance lost time	The remainder of the yellow time that can't be effectively used by traffic and all of the red clearance time.
Clearing distance	The distance traveled from the choice point to a point where the rear of the vehicle clears the far side of the intersection.
Compatible movements	Movements that can be served concurrently.
Concurrency group	Group of movements that can travel at the same time.
Conflicting movements	Movements that must be served in sequence.
Critical volume-to-capacity ratio	The proportion of available intersection capacity used by the critical movements.
Critical movement	A movement or sequence of movements that determine(s) the timing requirements for a concurrency group.
Cumulative vehicle diagram	A queuing diagram that represents the cumulative number of arrivals and departures over time at a signalized intersection.
Cycle length	"The time elapsed between the endings of two sequential terminations of a given interval." [HCM 2010]
D/D/1 model	A queuing model that assumes a deterministic arrival pattern, a deterministic service pattern, and one service channel.
Deterministic pattern	Arrival or departure flow patterns in which no randomness is involved.
Dilemma zone	A zone on the intersection approach where a vehicle can neither safely stop nor safely clear the intersection when the yellow is first displayed.
Displayed times	The times that a green, yellow, or red indication is displayed to the user.
Effective green ratio	"The ratio of the effective green time of a phase to the cycle length." [HCM 2010]
Effective green time	"The time during which a given traffic movement or set of movements may proceed at the saturation flow rate; it is equal to the split time minus the lost time." [HCM 2010]

Glossary

Term	Definition
Effective red time	"The time during which a given traffic movement or set of movements is directed to stop; it is equal to the cycle length minus the effective green time." [HCM 2010]
Flow profile diagram	A queuing diagram that represents the flow rates of the arrivals and departures at a signalized intersection.
Flow ratio	Ratio of the arrival flow rate to the saturation flow rate for a movement
Headway	The time between the passage of successive vehicles at a given point.
Intersection delay	"The total additional travel time experienced by drivers, passengers, or pedestrians as a result of control measures and interaction with other users of the facility, divided by the volume departing from the corresponding cross section of the facility." [HCM 2010]
Leading left turn	When left turns are served before the through movements.
Length of queue	The difference between the number of vehicles that have arrived at and departed from the intersection at any point in time.
Level of service	"A quantitative stratification of a performance measure or measures that represent quality of service, measured on an A–F scale, with LOS A representing the best operating conditions from the traveler's perspective and LOS F the worst." [HCM 2010]
Lost time	"The time during which a movement's phase is active and the approach is not used effectively by that movement; it is the sum of clearance lost time and start-up lost time." [HCM 2010]
Maximum queue length	The queue length at the end of the effective red.
Minimum cycle length	The time required to serve the critical movements as well as to account for the time that is lost due to the transition between phases.
Movement	"A term used to describe the user type (vehicle or pedestrian) and action (turning movement) taken at an intersection. Two different types of movements include those that have the right of way and those that must yield consistent with the rules of the road or the Uniform Vehicle Code." [STM]
Pedestrian clearance interval	The time required for a pedestrian to cross the intersection, assuming a walking speed of 3.5 ft/sec.
Permitted left turn	"A left … turn at a signalized intersection that is made by a vehicle during a time in the cycle in which the vehicle does not have the right-of-way." [HCM 2010]
Phase	"A controller timing unit associated with the control of one or more movements. The MUTCD defines a phase as the right-of-way, yellow change, and red clearance intervals in a cycle that are assigned to an independent traffic movement." [STM]
Pretimed control, Pretimed mode	"A signal control in which the cycle length, phase plan, and phase times are preset to repeat continuously." [HCM 2010]
Protected left turn	"The left … turns at a signalized intersection that are made by a vehicle during a time in the cycle when the vehicle has the right-of-way." [HCM 2010]
Quality of service	How well a transportation facility operates from the perspective of the users of the facility.
Queue	A line of vehicles that forms behind the vehicle in the server.
Queue accumulation polygon	A queuing diagram that represents the evolution of the queue length over time at a signalized intersection.

Term	Definition
Queue length	The number of vehicles in the queue at any given time.
Queue service time	The length of time required for the clearance of the queue.
Red clearance time	The time set for safe clearance of the intersection, or the travel time from the stop bar to when the rear of the vehicle clears the far side of the intersection.
Ring	A set of phases that operate in sequence.
Ring barrier diagram	"A graphical representation of phases within a set of rings and phases within a set of barriers." [STM]
Saturation flow rate	"The equivalent hourly rate at which previously queued vehicles can traverse an intersection approach under prevailing conditions, assuming that the green signal is available at all times and no lost times are experienced." [HCM 2010]
Saturation headway	The headway between vehicles in a departing queue at a signalized intersection.
Server	The first vehicle position at the stop bar in a queuing system.
Split phasing	When each approach at a signalized intersection is served in sequence.
Split proportion	The proportion of the cycle allocated to serve a phase.
Split time	The time allocated to serve a phase.
Startup lost time	The initial lag as drivers react to a change in the signal indication from red to green.
Stopping distance	The distance traveled while reacting to a stimulus and braking to a stop.
Timing stage	The interval during a signal cycle during which no displays change their indication.
Total delay	The delay experienced by all vehicles that arrive at and travel through the intersection.
Total uniform delay	The total delay experienced by vehicles during a cycle when the arrival flow is uniform.
Uniform arrivals or uniform flow	A flow pattern in which headways are uniform and constant.
Volume-to-capacity ratio	The ratio of the volume to the capacity of an intersection approach.
Yellow time	The duration that yellow is displayed and set so that a vehicle can travel from the choice point to the stop bar at constant speed.

Note: STM refers to reference [2]; HCM 2010 refers to reference [3].

PROBLEMS AND EXERCISES

Computational Problems

Representing Traffic Flow at Signalized Intersections (Section 2)

Problem 1. Prepare a flow profile diagram that represents the following conditions on one approach of a signalized intersection:

- Arrival flow rate = 600 veh/hr
- Saturation flow rate = 1900 veh/hr
- Queue service time = 13.8 sec
- Cycle length = 60 sec
- Green time = 30 sec
- Red time = 30 sec

Problem 2. Prepare a cumulative vehicle diagram that represents the following conditions on one approach of a signalized intersection:

- Vehicles arrive every 6 sec at a uniform rate.
- The cycle length is 60 sec, with red and green time intervals of 30 sec each.
- Vehicles depart every 2 sec after the beginning of green.
- The queue service time is 15 sec.

Problem 3. Consider the conditions given in Problem 2. Construct queue accumulation polygon that represents these conditions.

Problem 4. A flow profile diagram representing the arrival and departure patterns on one approach of a signalized intersection is shown in the figure below.

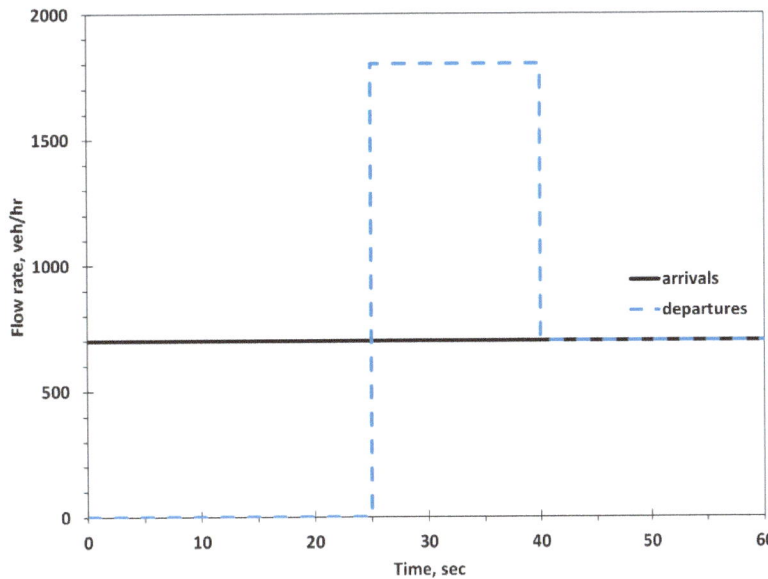

Based on the arrival and departure patterns shown in the figure, determine the:

- Arrival flow rate
- The saturation flow rate
- The queue service time
- The cycle length

Problem 5. Field data collected on one approach of a signalized intersection showed the following conditions:

- Cycle length = 72 sec
- Red interval = 30 sec
- Green interval = 42 sec
- Arrival flow rate = 900 veh/hr

- Saturation flow rate = 1800 veh/hr
- Vehicles depart every 2 sec

Prepare a cumulative vehicle diagram and a queue accumulation polygon that represent these conditions.

Problem 6. A cumulative vehicle diagram representing flow on one approach of a signalized intersection is given below. Interpret and explain the conditions represented in the diagram including the cumulative arrival and departure lines, the cycle length, and whether there is sufficient capacity to serve the demand.

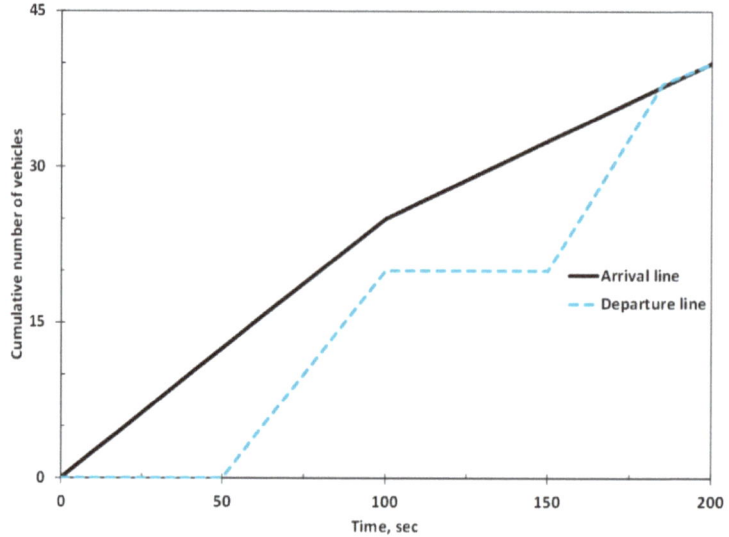

Problem 7. Consider the queue accumulation polygon given below. Interpret the conditions represented in the diagram. Find the arrival flow rate, the cycle length, red time, the number of vehicles that arrive on red (vr), the queue service time (g_s) and the arrival and departure rate. Construct the cumulative vehicle diagram, given a saturation flow rate of 1800 veh/hr.

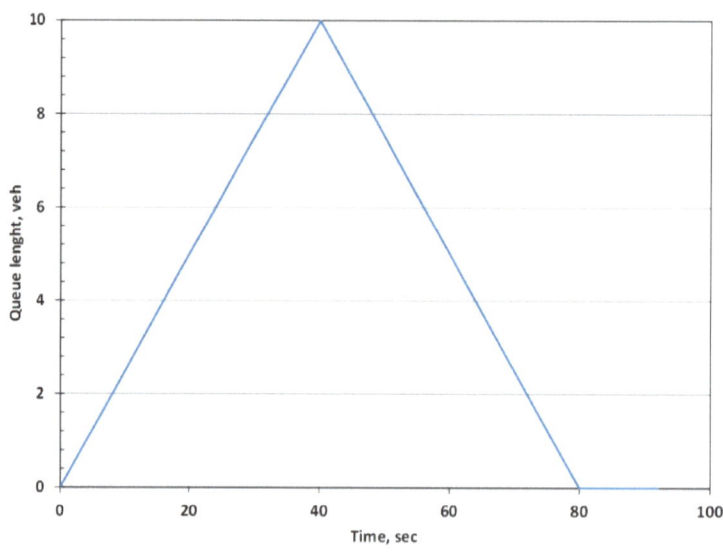

Sequencing and Controlling Movements (Section 3)

Problems and Exercises

Problem 8. Given the
intersection geometry,
movements, phases, and traffic
volumes as shown in the figure
at right:

- Determine the left turn
  phasing plan.
- Draw a ring barrier diagram
  that includes this left turn
  phasing plan (if protected
  left turns are required,
  assume leading).

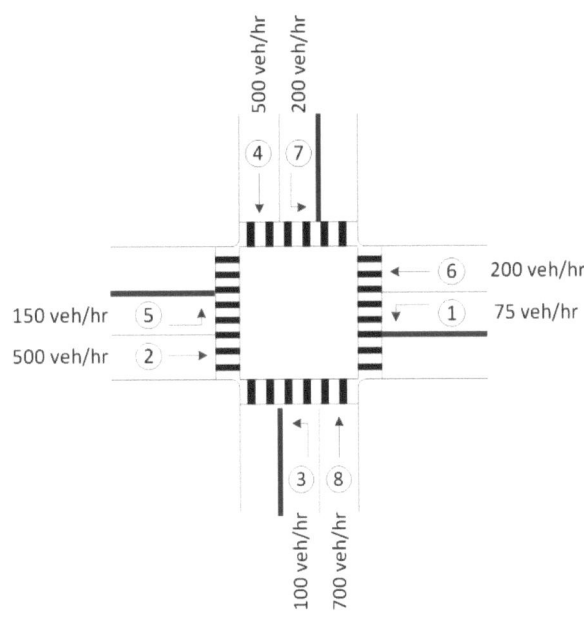

Problem 9. Draw a ring barrier
diagram showing the timing
stages given the timing data
shown in the table. Assuming
leading protected left turns.

Movement Number	Timing Required (sec)
1	10
2	25
3	15
4	35
5	5
6	15
7	15
8	30

Problem 10. A signalized intersection uses permitted left turn phasing for all four
left turn movements at a standard 4-leg intersection. Prepare a ring barrier
diagram that represents this phasing showing each phase and the movement or
movements that it controls, the sequence of the phases in each of the two rings,
and the locations of the barriers that divide the two concurrency groups.

Problem 11. A signalized intersection uses the leading protected left turn phasing
for the east-west concurrency group and lagging protected left turn phasing for
the north-south concurrency group. Prepare a ring barrier diagram that
represents this phasing showing each phase and the movement or movements
that it controls, the sequence of the phases in each of the two rings, and the
locations of the barriers that divide the two concurrency groups.

Problem 12. Prepare a ring barrier diagram that shows the timing stages and duration of each stage given the following conditions:

- Leading protected left turns for the east-west concurrency group
- Permitted left turns for the north-south concurrency group

Movement Number	Timing Required (seconds)
1	5
2	25
3	10
4	30
5	15
6	40
7	10
8	25

Problem 13. Draw a ring-barrier diagram for the following conditions:

- Leading protected LT for NS concurrency group
- Permitted LT for EW concurrency group

Show the resulting timing stages in the diagram based on the data given in the table on the right.

Phase	Timing Required (sec)
1	20
2	30
3	10
4	25
5	15
6	25
7	15
8	20

Problem 14. The geometry for a signalized intersection is given as follows:

- The EB and WB approaches have one through lane and one left turn lane
- The NB and SB approaches have one lane for both the through and left turn movements.

Movement	Flow rate (veh/hr)
NBLT	100
SBTH	400
SBLT	150
NBTH	300
EBLT	100
WBTH	700
WBLT	150
EBTH	650

The flow rate for each movement and the required phase durations are given in the tables at right.

Phase	1	2	4	5	6	8
Duration	15	30	25	10	35	30

Based on these flow rates and intersection geometry, what left turn phasing would you recommend for each of the four approaches? Create a ring-barrier diagram for this intersection. Construct a ring-barrier diagram for the intersection showing the resulting timing stages.

Problem 15. The geometry of a T-intersection includes:
- One through lane for EB and WB
- Left turn lane for WB
- Right turn bay for EB
- One left turn lane and one right turn bay for NB

Movement	Flow Rate (veh/hr)
NBLT	300
NBRT	200
WBTH	700
WBLT	150
EBTH	300
EBRT	50

The flow rate for each movement and the phase durations are given in the tables on the right. Create ring-barrier diagram for this case. Construct ring-barrier diagram for the intersection showing the resulting timing stages.

Phase	1	2	3	6
Duration	10	25	20	35

Yellow and Red Clearance Intervals (Section 4)

Problem 16. Determine Y and RC intervals for all four approaches for the signalized intersections represented in the sketch below. Assume:
- δ = 1.0 sec
- a = 10 ft/sec²
- L_v = 20 ft

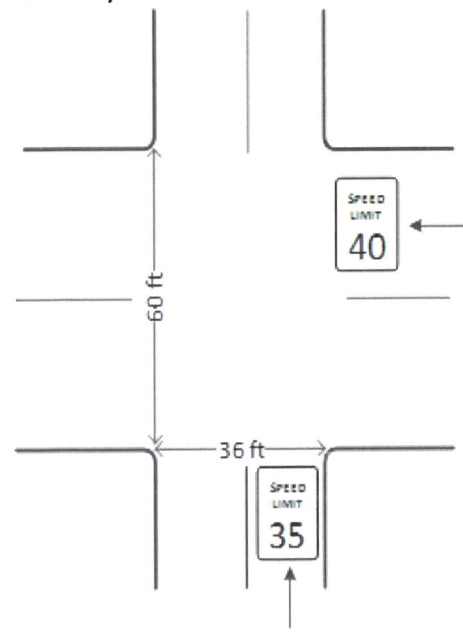

Problem 17. If the speed on the NS approaches for Problem 16 is changed to 40 mi/hr, and Y and RC are not changed, what is the implication for the safe operation of the intersection? Back up conclusions with supporting calculations.

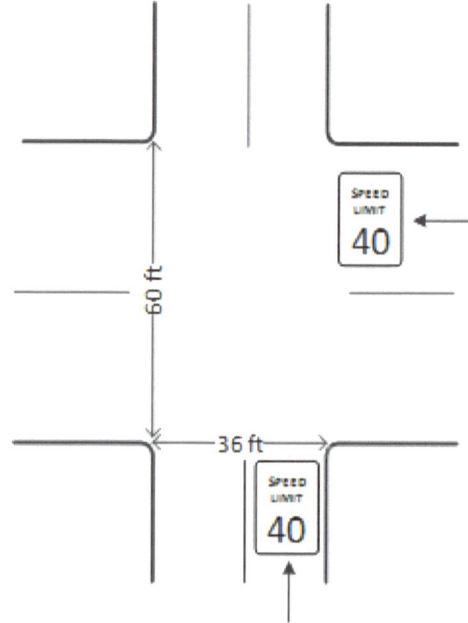

Problem 18. Assume the data and results given in Example 7 (page 36) of this module. Suppose that, despite your calculations, the city engineer instructs you to reduce the yellow time from 3.6 sec to 3.0 seconds (with the red clearance time unchanged at 1.2 seconds). Show that this difference results in a dilemma zone for some drivers approaching the intersection when yellow is first displayed.

Problem 19. Consider an intersection with a speed limit of 30 mi/hr. If a vehicle is located 82 feet upstream of the stop bar when yellow is first displayed, does a dilemma zone exist for this driver? Assume:
- $\delta = 1.0$ sec
- $a = 10$ ft/sec^2
- $L_v = 20$ ft
- $w = 80$ ft

Problem 20. Consider an intersection with a speed limit of 30 mi/hr. What is the closest point upstream of the stop bar that a vehicle can safely stop? What is the farthest point upstream of the stop bar that a vehicle can safely clear the intersection? Common on the Y and RC times used at this intersection.   Assume:
- $\delta = 1.0$ sec
- $a = 10$ ft/sec^2
- $L_v = 20$ ft
- $w = 80$ ft
- $Y + RC = 4.0$ sec

Capacity (Section 5)

Problem 21. For one approach of a signalized intersection, the saturation headway is 2 sec/veh. The green is displayed for 20 sec, while the sum of the yellow and red clearance displays is 5 sec. The lost time is 4 sec and the cycle length is 65 sec. What is the capacity of the approach?

Sufficiency of Capacity (Section 6)

Problem 22. Suppose that the volume on one approach of a signalized intersection is 800 veh/hr and the saturation flow rate for the approach is 1900 veh/hr. Assume the effective green ratio to be 0.32. Does this effective green ratio provide sufficient capacity? If not what can be done to provide sufficient capacity on this approach?

Problem 23. Determine the sufficiency of capacity for the intersection and data given below and in the figure at right:

- C = 60 sec
- s = 1900 veh/hr/lane for all movements
- L = 4 sec/phase
- Leading protected LTs

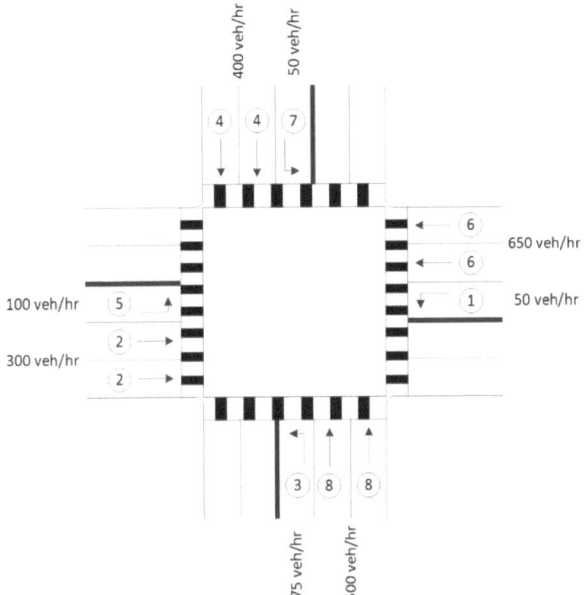

Include the following steps in your calculations:

- Compute Y for each movement
- Determine the flow ratio sums for each concurrency group
- Within each concurrency group, identify the sequence of movements with the highest flow ratio sum
- Determine X_c for the intersection
- Determine the sufficiency of capacity rating

Problem 24. Given the traffic volume data and lane configuration shown in the figure. Based on the critical degree of saturation, would you recommend protected or permitted LT phasing?

Assume:

- Saturation flow rates of 1900 veh/hr/lane for protected LT or TH movements, or 450 veh/hr/lane for permitted LT movements.
- Cycle length = 60 sec
- Lost time = 4 sec/phase

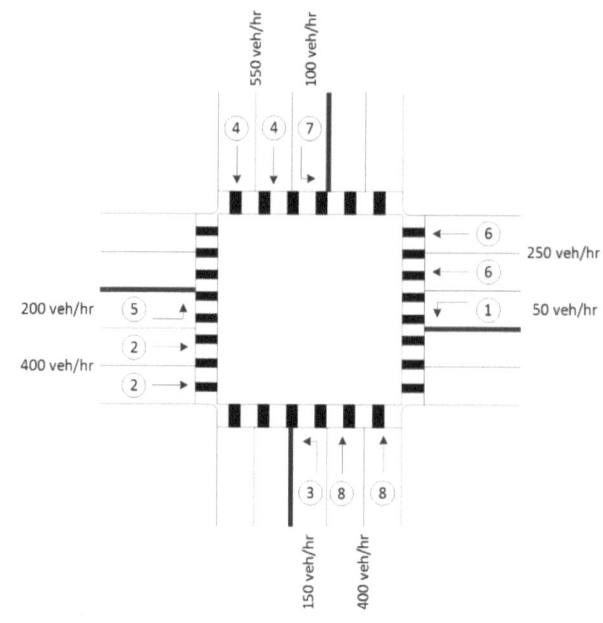

Problem 25. Given the traffic volume data and lane configuration shown in the figure. Based on the critical degree of saturation, would you recommend protected or permitted LT phasing?

Assume:

- Saturation flow rates of 1900 veh/hr/lane for protected LT or TH movements, or 450 veh/hr/lane for permitted LT movements.
- Cycle length = 45 sec
- Lost time = 4 sec/phase

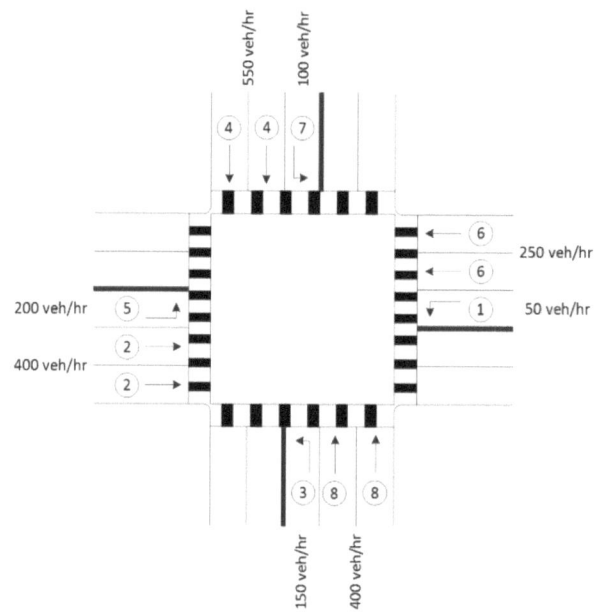

Problem 26. Given the traffic volume data and lane configuration shown in the figure. Assuming protected left turns for the east-west movements and permitted left turns for the north-south movements, is there sufficient capacity to accommodate the traffic demand? Assume a 70 second cycle length, with saturation flow rates of 1900 veh/hr/lane for protected LT or TH movements, or 450 veh/hr/lane for permitted LT movements.

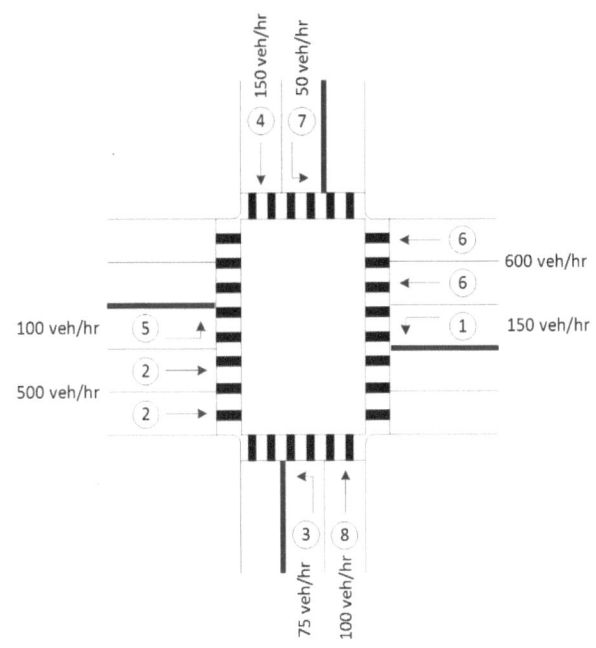

Problem 27. Given traffic volume data and the lane configurations shown in the figure at right. Determine the critical degrees of saturation for each concurrency group given the data below. Assume leading protected left turns and an 80 sec cycle length.

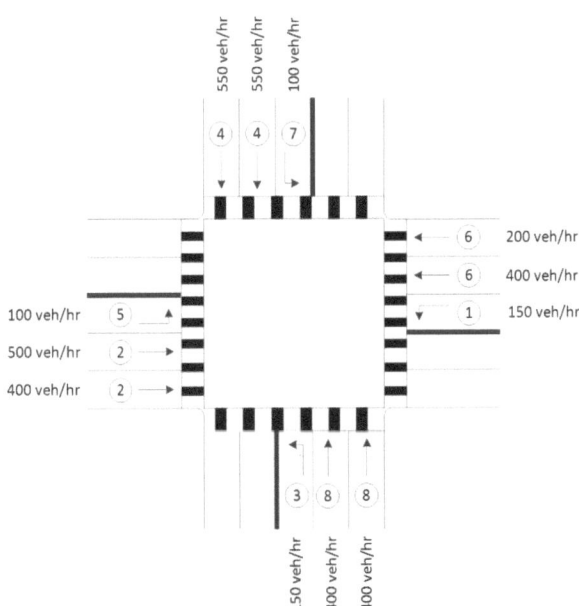

Problem 28. The following table (in the form of a ring barrier diagram), shows the flow ratio for each of the eight movements for a phasing plan based on leading protected left turns. Using these flow ratios, construct a diagram showing the six timing stages that result and the relative proportion of the hour required to serve each stage.

	East-West Concurrency Group			North-South Concurrency Group		
Ring 1	Ø1	.10	Ø2	.35	Ø3	.15	Ø4	.25
Ring 2	Ø5	.15	Ø6	.30	Ø7	.20	Ø8	.20

How does Y affect signal timing (or the time required to serve a given movement)? What is a timing stage? How does the sequencing implied in two ring operation provide the opportunity for more efficiency than if the left turns are served first, followed by the TH movements?

Problem 29. A standard four-approach intersection has the geometric and volume characteristics shown in the figure at right. The phasing plan is also shown. The cycle length is 90 sec and the lost time is 4 sec per phase. The saturation flow rate is 1900 veh/hr for through movements and protected left turn movements and is 450 veh/hr for permitted left turn movements. Determine the sufficiency of capacity for the intersection.

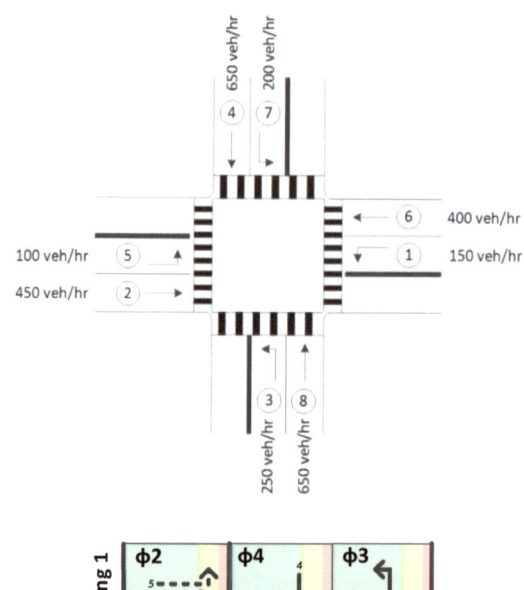

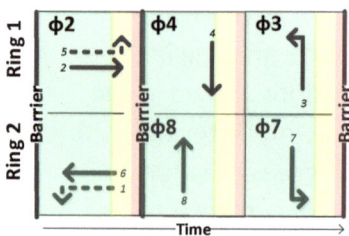

Delay and Level of Service (Section 7)

Problem 30. Consider an approach at a signalized intersection with the following conditions:

- Saturation flow rate = 1900 veh/hr
- Cycle length = 60 sec
- Effective green ratio = 0.5
- During the first cycle, the uniform flow rate is 1000 veh/hr. In the following cycles, the flow rate drops to 300 veh/hr.

Draw the flow profile diagram, the cumulative vehicle diagram, and the queue accumulation polygon for the conditions described above. Calculate the time after the beginning of the first cycle when the queue clears.

Problem 31: Prepare a flow profile diagram, a cumulative vehicle diagram, and a queue accumulation polygon for an arrival pattern consisting of arrivals on green only for one approach of a signalized intersection (and no arrival flow at any other time). Label each of the three diagrams using the variables listed below, as well as the variables that are represented on the x and y axes. Assume:

- Cycle length = C
- Effective red = r
- Effective green = g
- Saturation flow rate = s
- Queue service time = g_s

Problem 32: Prepare a flow profile diagram, a cumulative vehicle diagram, and a queue accumulation polygon for an arrival pattern consisting of arrivals on red only for one approach of a signalized intersection (and no arrival flow at any other time). Label each of the three diagrams using the variables listed below, as well as the variables that are represented on the x and y axes. Assume:

- Cycle length = C
- Effective red = r
- Effective green = g

- Saturation flow rate = s
- Queue service time = g_s

Problem 33: Prepare a flow profile diagram, a cumulative vehicle diagram, and a queue accumulation polygon for an arrival pattern consisting of uniform arrivals beginning half way through the red interval and ending half way through the green interval for one approach of a signalized intersection. The arrival rate equals zero at all other times. Label each of the three diagrams using the variables listed below, as well as the variables that are represented on the x and y axes. Assume:

- Cycle
- length = C
- Effective red = r
- Effective green = g

- Saturation flow rate = s
- Queue service time = g_s

Problem 34: Given the following conditions for one approach of a fixed time signalized intersection.

Given data	Value	Units
average arrival rate, v	330	veh/hr
rate during red, v_r	400	veh/hr
rate during green, v_g	250	veh/hr
C	75	sec
r	40	sec
g	35	sec
g/C	0.467	
r/C	0.533	
s	1900	veh/hr

Based on the data given above, answer the following questions, showing the calculations that support your answer:

- What is the rate of queue formation?
- What is the rate of queue clearance?
- What is the maximum queue length (maximum number of vehicles in the queue)?
- What is the queue service time?
- Based on the given data and your results from questions 1 through 4, prepare a queue accumulation polygon for a period of one cycle.
- What is the total number of vehicles that arrive during one cycle?
- What is the total delay?
- What is the average delay per vehicle?

Problem 35: An approach to a pretimed signalized intersection has a saturation flow rate of 1700 veh/hr. The cycle length is 60 sec and the effective red is 40 sec. During three consecutive cycles 15, 8, and 4 vehicles arrive. The arrival pattern should be assumed to be uniform during each cycle.
- Prepare a flow profile diagram, a cumulative vehicle diagram, and a queue accumulation polygon for these conditions.
- Determine the total vehicle delay and the average delay per vehicle for each cycle, and for all three cycles.
- For the queue present at the beginning of each of the three green intervals, how long would it take for each queue to clear?
- Is there sufficient capacity on capacity on this approach to serve the demand?

Problem 36. An intersection approach has an arrival rate of 720 veh/hr and a saturation flow rate of 1800 veh/hr. The cycle length is 100 sec, the effective red is 50 sec and the effective green is 50 sec. Determine the queue service time, the average delay and the total delay for this approach. Also, construct a cumulative vehicle diagram and queue accumulation polygon.

Cycle Length and Split Times (Section 8)
Problem 37. The flow ratios for each movement (shown in ring barrier format) are given below. Protected LT phasing is assumed. Assume lost time per phase of 4 sec. Compute the cycle length.

Ring 1	Y_1	.10	Y_2	.25	Y_3	.15	Y_4	.25
Ring 2	Y_5	.10	Y_6	.20	Y_7	.10	Y_8	.255

Problem 38. Determine the left turn phasing, the cycle length and the split times given the following data:
- Pretimed signalized intersection
- Lost time per phase = 4 sec
- Acceptable cycle length range = 60-90 sec
- Minimum split time = 10 sec
- Saturation flow rates = 1900 veh/hr/lane for protected LTs and TH movements, and 450 veh/hr/lane for permitted LTs.
- One TH lane and one exclusive LT lane on each approach
- Arrival flow rates as noted in the table below

Ring 1	v_1	100	v_2	500	v_3	175	v_4	325
Ring 2	v_5	75	v_6	450	v_7	125	v_8	375

Problem 39. Determine the cycle length and split times given the following data:
- Pretimed signalized intersection that operates with leading protected left turns
- Lost time per phase = 4 sec
- Acceptable cycle length range = 60-90 sec
- Minimum split time = 10 sec
- Arrival flow rates and saturation flow rates as noted in the table below

Ring 1	V_1	250	V_2	500	V_3	125	V_4	300
	S_1	1900	S_2	1900	S_3	1900	S_4	1900
Ring 2	V_5	75	V_6	425	V_7	75	V_8	600
	S_5	1900	S_6	1900	S_7	1900	S_8	1900

Problem 40. Determine the cycle length and split times given the following data:
- Pretimed signalized intersection that operates with leading protected left turns
- Lost time per phase = 4 sec
- Acceptable cycle length range = 60-90 sec
- Minimum split time = 10 sec
- Arrival flow rates and saturation flow rates as noted in the table below

Ring 1	V_1	100	V_2	350	V_3	100	V_4	400
	S_1	1900	S_2	1900	S_3	1900	S_4	1900
Ring 2	V_5	50	V_6	425	V_7	125	V_8	400
	S_5	1900	S_6	1900	S_7	1900	S_8	1900

Problem 41. Determine the left turn phasing, the cycle length and the split times given the following data:
- Pretimed signalized intersection
- Lost time per phase = 4 sec
- Acceptable cycle length range = 60-90 sec
- Minimum split time = 10 sec
- Saturation flow rates = 1900 veh/hr/lane for protected LTs and TH movements, and 450 veh/hr/lane for permitted LTs.
- One TH lane and one exclusive LT lane on each approach
- Arrival flow rates as noted in the table below

Ring 1	V_1	100	V_2	525	V_3	50	V_4	325
Ring 2	V_5	150	V_6	650	V_7	125	V_8	300

Problem 42. Determine the left turn phasing, the cycle length and the split times given the following data:

- Pretimed signalized intersection
- Lost time per phase = 4 sec
- Acceptable cycle length range = 60-90 sec
- Minimum split time = 10 sec
- Saturation flow rates = 1900 veh/hr/lane for protected LTs and TH movements, and 450 veh/hr/lane for permitted LTs.
- One TH lane and one exclusive LT lane on each approach
- Arrival flow rates as noted in the table below

Ring 1	V_1	100	V_2	525	V_3	175	V_4	450
Ring 2	V_5	150	V_6	650	V_7	150	V_8	500

Problem 43. The flow ratios for an intersection have been computed and are given in the table below. The critical flow ratio sums for the east-west and north-south concurrency groups are .386 and .424, respectively. The total lost time per cycle is 16 sec. What value would you recommend for the cycle length? Assume protected left turn phasing.

	East-West Concurrency Group		North-South Concurrency Group	
Ring 1	Y_1=.125	Y_2=.253	Y_3=.174	Y_4=.25
	Y_{EW1} = .378		Y_{NS1} = .424	
Ring 2	Y_5=.15	Y_6=.236	Y_7=.123	Y_8=.285
	Y_{EW2} = .386		Y_{NS2} = .408	

Problem 44.
Use the data given in problem 43 data and calculate the split times for the critical phases.

Signal Timing (Section 9)

Problem 45. Determine signal timing and performance given the following data about four leg pretimed signalized intersection. Assume that the minimum green time is 5 sec.

	East-West Concurrency Group				North-South Concurrency Group			
Ring 1	v_1	125	v_2	375	v_3	75	v_4	625
Ring 2	v_5	75	v_6	575	v_7	175	v_8	275

	EW Approaches		NS Approaches	
Speed	25	mi/hr	25	mi/hr
Intersection width	42	ft	42	ft
Number of TH approach lanes	1		1	
Vehicle length		20	ft	

Problem 46. Determine signal timing and performance given the following data about four leg pretimed signalized intersection. Assume that the minimum green time is 5 sec.

	East-West Concurrency Group				North-South Concurrency Group			
Ring 1	v_1	250	v_2	500	v_3	175	v_4	325
Ring 2	v_5	250	v_6	475	v_7	300	v_8	450

	EW Approaches		NS Approaches	
Speed	25	mi/hr	25	mi/hr
Intersection width	42	ft	42	ft
Number of approach lanes	1		1	
Vehicle length		20	ft	

Problem 47. Determine signal timing and performance given the following data about four leg pretimed signalized intersection. Assume that the minimum green time is 5 sec.

	East-West Concurrency Group				North-South Concurrency Group			
Ring 1	v_1	125	v_2	750	v_3	125	v_4	625
Ring 2	v_5	175	v_6	675	v_7	75	v_8	600

	EW Approaches		NS Approaches	
Speed	25	mi/hr	25	mi/hr
Intersection width	66	ft	66	ft
Number of approach lanes	2		2	
Vehicle length		20	ft	

Problem 48. Determine signal timing and performance given the following data about four leg pretimed signalized intersection. Assume that the minimum green time is 5 sec.

	East-West Concurrency Group				North-South Concurrency Group			
Ring 1	v_1	150	v_2	500	v_3	150	v_4	350
Ring 2	v_5	125	v_6	425	v_7	150	v_8	600

	EW Approaches		NS Approaches	
Speed	25	mi/hr	25	mi/hr
Intersection width	42	ft	42	ft
Number of approach lanes	1		1	
Vehicle length		20	ft	

Problem 49. A standard 4-approach intersection has the geometric and volume characteristics shown the figure below. The saturation flow rate is 1900 veh/hr/lane for through movements and protected left turn movements and the saturation flow rate is 450 veh/hr/lane for permitted left turn movements. The speed limit is 35 mi/hr. The lost time is 4 sec/phase. Assume a perception-reaction time of 1 sec and deceleration rate of 10 ft/sec^2. Assume also a vehicle length equal to 20 ft.

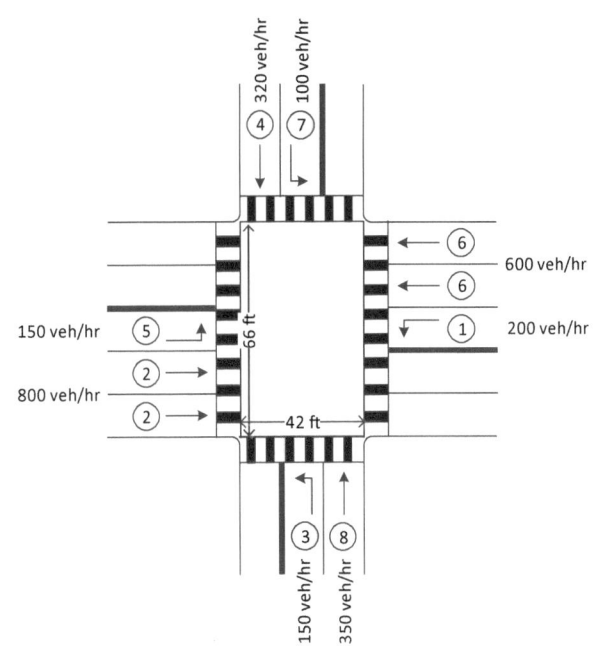

Develop a signal phasing and timing plan for this intersection.

Problem 50. Determine signal timing and performance given the following data about four leg pretimed signalized intersection. Assume that the minimum green time is 5 sec.

Ring 1	v_1	175	v_2	900	v_3	250	v_4	975
	s_1	1900	s_2	3800	s_3	450	s_4	1900
Ring 2	v_5	150	v_6	750	v_7	225	v_8	1200
	s_5	1900	s_6	3800	s_7	450	s_8	1900

	EW Approaches		NS Approaches	
Speed	35	mi/hr	35	mi/hr
Intersection width	66	ft	66	ft
Number of approach lanes	2		2	
Vehicle length	20	ft		

Problem 51. Determine signal timing and performance given the following data about four leg pretimed signalized intersection. Assume that the minimum green time is 5 sec.

Ring 1	V_1	150	V_2	825	V_3	150	V_4	475
	S_1	1900	S_2	3800	S_3	1900	S_4	1900
Ring 2	V_5	190	V_6	1000	V_7	180	V_8	500
	S_5	1900	S_6	3800	S_7	1900	S_8	1900

	EW Approaches		NS Approaches	
Speed	35	mi/hr	35	mi/hr
Intersection width	66	ft	42	ft
Number of approach lanes	2		1	
Vehicle length			20	ft

Field Exercises

Exercise #1: Introduction to Signalized Intersections: Intersection Elements

Purpose

The purpose of this exercise is to observe the key elements of a signalized intersection in the field.

Tasks

1. Locate, sketch, and/or label the following elements for the intersection to which you've been assigned: lanes, detectors, shoulders, crosswalks, stop lines, pavement markings, vehicle signal heads, pedestrian signal heads, signal cabinet, and poles.

2. Observe the operation of the intersection for three cycles (service to each approach three times). Note whether or not the queue clears before the end of green or if there are still vehicles waiting to be served when yellow is first displayed. Are there pedestrians crossing the intersection? Are there special signal displays such as flashing yellow arrows or solid green arrows? Is there a pedestrian countdown timer?

Deliverables

Prepare a two page summary of your observations including the sketch that you make.

Exercise #2: Queuing Systems: Field Observations

Purpose. The purpose of this activity is to observe the process of queuing at a signalized intersection and to learn how to prepare a cumulative vehicle diagram with the field data that you collect.

Tasks
[Note: All times should be recorded to the nearest second.]

1. For the intersection approach that you have been assigned, watch the traffic flow for several cycles. Observe the queue and how far back from the stop bar the queue reaches. The maximum extent of the queue is the furthest point the line of cars reaches back from the stop bar, as shown in the Figure 73. This location is considered to be the entry point into the system. The time at which a vehicle crosses this point will be considered the time it enters the system. The stop bar, at the edge of the intersection, is considered the exit point of the system. When a car crosses this line it will be considered as exiting the system.
2. Record the times that each vehicle arrives into the system and leaves the system for **three** cycles for **one** lane. An example of the results of the data collection for one cycle is shown in Table 23. A blank field data collection form is given in Table 24.
3. As you collect the arrival and departure times, also record the times that the signal indication turns red and green for each of the three cycles.
4. Based on the field data that you collect, calculate the arrival and departure volumes of this lane for each cycle.
5. Plot the arrivals and departures of vehicles for each cycle. The resulting diagram is a cumulative vehicle diagram. Observe the variation in queue length between cycles and the variation in delay between individual vehicles. An example cumulative vehicle diagram based on the data in Table 23 is shown in Figure 74.
6. Were all of the observed cycle lengths the same? Why do you think they are/aren't?
7. What differences did you notice between the cumulative vehicle diagram constructed from your field data and the same diagram for the D/D/1 queuing model presented earlier in this module?

Deliverables
Prepare a two page report summarizing your observations and conclusions, as well as the data that you collect, the chart that you prepare, and answers to the questions listed above.

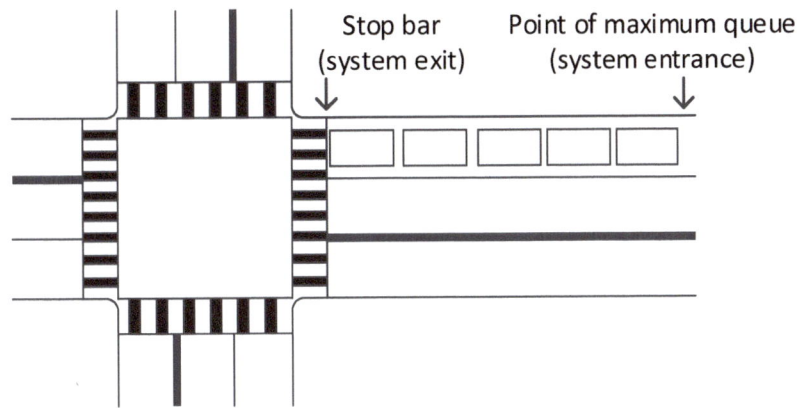

Figure 73. Maximum extent of queue

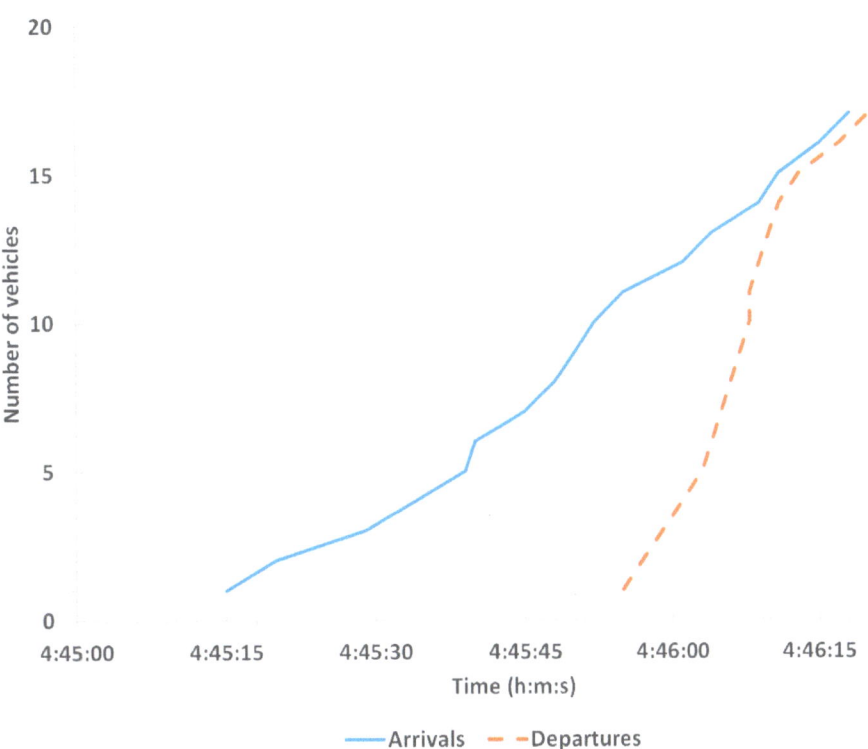

Figure 74. Cumulative vehicle diagram

Table 23. Example field data

	Cycle 1		Cycle 2		Cycle 3	
		Time		Time		Time
	red	4:45:14	red		red	
	green	4:45:54	green		green	
	Arrival Time	Departure Time	Arrival Time	Departure Time	Arrival Time	Departure Time
1	4:45:15	4:45:55				
2	4:45:20	4:45:57				
3	4:45:29	4:45:59				
4	4:45:34	4:46:01				
5	4:45:39	4:46:03				
6	4:45:40	4:46:04				
7	4:45:45	4:46:05				
8	4:45:48	4:46:06				
9	4:45:50	4:46:07				
10	4:45:52	4:46:08				
11	4:45:55	4:46:08				
12	4:46:01	4:46:09				
13	4:46:04	4:46:10				
14	4:46:09	4:46:11				
15	4:46:11	4:46:13				
16	4:46:15	4:46:17				
17	4:46:18	4:46:20				
18						

Problems and Exercises

Table 24. Field data collection form

	Cycle 1			Cycle 2			Cycle 3	
		Time			Time			Time
	red			red			red	
	green			green			green	
	Arrival Time	Departure Time		Arrival Time	Departure Time		Arrival Time	Departure Time
1								
2								
3								
4								
5								
6								
7								
8								
9								
10								
11								
12								
13								
14								
15								
16								
17								
18								

REFERENCES

[1] Federal Highway Administration. *Manual of Uniform Traffic Control Devices for Streets and Highways.* Washington, D.C., 2009.

[2] Urbanik, T., A. Tanaka, B. Lozner, E. Lindstrom, K. Lee, D. Gettman, S. Sunkari, K. Balke, D. Bullock. Signal Timing Manual. National Cooperative Research Program 03-103, Washington, D.C., 2015.

[3] Transportation Research Board. *Highway Capacity Manual*. National Research Council, Washington D.C., 2010.

[4] Kyte, Michael and Tom Urbanik. *Traffic Signal Systems Operations and Design: An Activity-Based Learning Approach* (Book 1: Isolated Intersections). Pacific Crest Software, Inc., 2012.

[5] Pignataro, Louis J. *Traffic Engineering Theory and Practice*. Prentice-Hall, 1973.

INDEX

[**Bold** numbers indicate the page where the definition is given.]

www.ingramcontent.com/pod-product-compliance
Lightning Source LLC
Chambersburg PA
CBHW050718180526
45159CB00003B/1061